Differential Thermal Analysis

Differential Thermal Analysis

A GUIDE TO THE TECHNIQUE AND ITS APPLICATIONS

M. I. POPE

*Principal Lecturer in Physical Chemistry,
Portsmouth Polytechnic, Portsmouth, U.K.*

M. D. JUDD

*European Space Research and Technology Centre
Noordwijk, Netherlands.*

London · Bellmawr, N.J. · Rheine

Heyden & Son Ltd., Spectrum House, Hillview Gardens, London NW4 2JQ, U.K.
Heyden & Son Inc., Kor-Center East, Bellmawr, N.J. 08030, U.S.A.
Heyden & Son GmbH, Münsterstrasse 22, 4440 Rheine/Westf., Germany

ISBN 0 85501 229 3

Printed litho in Great Britain by W & J Mackay Ltd, Chatham

Contents

Preface

This book is intended to show the engineer, scientist, or technologist, how differential thermal analysis can be used to solve problems encountered in his own particular field of work. Numerous text books already exist which are concerned primarily with the technique and its theoretical interpretation, or with a single specialized topic, such as the use of differential thermal analysis (DTA) in plastics technology. Accordingly, a detailed mathematical approach has deliberately been avoided; mathematics are only introduced to an extent necessary to interpret DTA curves which may be obtained. Any further information required can generally be obtained from the references cited.

The authors believe that there are a great many instances where DTA could have provided a quick and clear answer to a problem, but was not employed because the technique was unfamiliar to the persons concerned. We have therefore set out to describe how DTA has proved to be of value in various fields, some of which are well known, and others which have previously been confined to a very small group of workers.

Portsmouth
July 1977

M. I. Pope
M. D. Judd

CHAPTER 1
Introduction

Differential thermal analysis (DTA) is a technique which involves heating (or cooling) a test sample and an inert reference sample under identical conditions and recording any temperature difference which develops between them. This differential temperature is then plotted either against time, or against the temperature at some fixed point within the apparatus. Any physical or chemical change occurring to the test sample which involves the evolution of heat will cause its temperature to rise temporarily above that of the reference sample, thus giving rise to an exothermic peak on the DTA plot. Conversely, a process which is accompanied by the absorption of heat will cause the temperature of the test sample to lag behind that of the reference material, leading to an endothermic peak.

However, even where no physical or chemical process is occurring, a small and steady differential temperature normally develops between the test and reference materials. This is due primarily to differences in the heat capacity and thermal conductivity of the two materials, but is also influenced by many other factors, such as sample mass and packing density. It therefore follows that DTA can be used to study transitions in which no heat is evolved or absorbed by the sample, as may be the case in certain solid–solid phase changes. The difference in the heat capacity of the sample before and after the transition has occurred will then be reflected in a new steady differential temperature being established between the test and reference samples. The base line of the differential thermal curve will accordingly show a sudden discontinuity at the transition temperature, while the slopes of the curve in the regions above and below this temperature will usually differ significantly.

ORIGIN AND DEVELOPMENT OF DTA

The discovery of DTA is usually credited to H. Le Châtelier in 1887, following the publication of two papers entitled, 'On the action of heat on clays'[1] and 'On the constitution of clays',[2] in *Comptes Rendus*, of the French Academy of Sciences. Much the same work was also reported in the Bulletin of the French Mineralogical Society,[3] again in French, while a German version appeared simultaneously in *Zeitschrift für Physikalische Chemie*.[4]

1

Le Châtelier did not in fact measure the difference in temperature between test and reference materials. His technique involved immersing the junction of a Pt-Pt 10% Rh thermocouple in clay, packed into a 5 mm platinum crucible. This in turn was placed in a larger crucible and surrounded with calcined magnesia, the whole assembly then being placed in a furnace. A heating rate of 4 °C every two seconds was employed and the thermocouple e.m.f. recorded photographically at regular intervals of time. Hence when no changes were occurring in the clay sample, his photographic record consisted of a series of evenly spaced lines, as the temperature rose steadily. Endothermic dehydration of the clay caused its temperature to lag behind that of the furnace, giving rise to an increase in the line spacing until the process was complete. Le Châtelier also noted that in certain places on his photographic records, the lines bunched together, corresponding to evolution of heat by the clay sample. The clays studied included halloysite, allophane, kaolin, pyrophyllite and montmorillonite. In each case, Le Châtelier concluded that the clays could be classified according to their heating patterns (see Fig. 1.1).

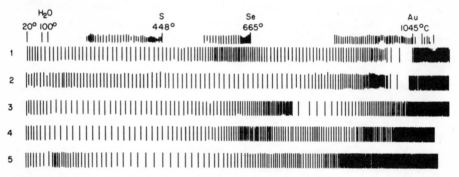

Fig. 1.1. The original photographic record of heating curve data published by H. Le Châtelier in 1887 (see Ref. 1). The uppermost record represents calibration of the apparatus from the boiling points of water, sulphur and selenium, and the melting point of gold. The lower records are for the clays (1) halloysite (2) allophane (3) kaolin (4) pyrophyllite and (5) montmorillonite,

This remarkable French scientist also employed the now common practice of calibrating his apparatus by means of the melting or boiling points of substances obtainable in a high state of purity. He actually used the boiling point of water at atmospheric pressure (100 °C), sulphur (448 °C), selenium (665° C), and the melting point of gold (1045 °C), the temperatures in parentheses being the values accepted at that time.

On this evidence, there seems little doubt but that Le Châtelier can justly be regarded as the originator of the technique of DTA. Nevertheless, because his method was not strictly differential, it lacked sensitivity. It was not until twelve years later that Sir W. C. Roberts-Austen published a description of the apparatus which forms the essential basis of all modern differential thermal analysers.

The 'Fifth report to the alloys research committee: steel', appeared in 1899, in both the *Proceedings of the Institution of Mechanical Engineers*[5] and in *The Metallographist*.[6] Differential thermal analysis was carried out using two thermocouples connected in opposition, the output signal being indicated on a sensitive mirror galvanometer. The steel sample to be studied was in the form of a cylindrical block, drilled through its centre so that one thermocouple junction could be inserted within it. The reference sample was identical in size and consisted either of copper–aluminium alloy or of fireclay, depending on the temperature range to be studied. Into it was inserted the second thermocouple junction. Both blocks were then placed in a tubular furnace, which was subsequently evacuated. A second, less sensitive galvanometer was used to record the temperature of the reference sample. A circuit diagram of Robert-Austen's apparatus is shown in Fig. 1.2 and a typical DTA curve obtained by him, on cooling a sample of

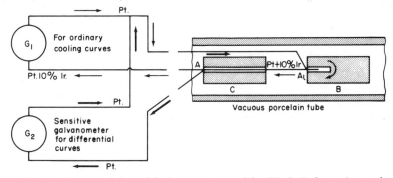

Fig. 1.2. Circuit diagram of the original apparatus used by W. C. Roberts–Austen in 1899 (see Ref 6). The galvanometer G_1 records the temperature of the reference sample and G_2 records the differential temperature.

'electro-iron', in Fig. 1.3. The onset temperatures of the exothermic peaks were noted as being 1132 °C, 895 °C, 766 °C, with broad exotherms occurring between 600 and 550 °C and between 500 and 450 °C, then a final sharp peak at 261 °C. This technique was used to determine the phase diagram for carbon steels and hence to study the properties of sections of railway line from various sources and manufactured in differing ways.

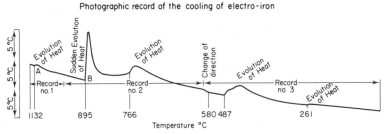

Fig. 1.3. Differential thermal analysis curve obtained by W. C. Roberts–Austen in 1899 (see Refs 5 and 6) for 'electro-iron', using platinum as the reference material.

That the experimental procedure, and method of presenting and interpreting data, has changed so little in the following three quarters of a century, is a striking tribute to the outstanding work of Roberts-Austen. The improvements in technique which have come about since his time are due almost entirely to electronic methods of temperature control and in the processing and recording of the thermocouple signals.

In the four decades following the publication of Roberts-Austen's work, sporadic reports appear in various journals, generally reporting to use of DTA to characterize minerals and inorganic compounds. Prominent among these are papers by Kurnakov,[7] by Orcel and Mlle Caillère[8] and by Berg[9] and his co-workers. During this period, it was progressively established that a DTA curve was characteristic of a substance and could be used as a means of identification. The technique proved of particular value in the study of clays[8] and silicaceous minerals, where the structural similarity between different materials rendered the interpretation of X-ray diffraction data extremely difficult.

By 1948, it had been established that the area under a DTA peak gives a direct and quantitative measure of the amount of heat evolved or absorbed in the corresponding process; furthermore that the area is not affected by the heat capacity of the sample. Kerr and Kulp[10] put forward a theoretical interpretation for the area and shape of a DTA peak, but did not take account of temperature gradients within the sample itself.

Controversy continued as to whether the significant point on a DTA peak was the point at which the curve first showed a detectable deviation from the original base line, or the peak temperature itself. Frederickson[11] preferred the onset temperature and suggested that it could best be determined from the first derivative of the DTA curve. Conversely, Morita and Rice,[12] in a study of a wide range of organic compounds including both natural and synthetic polymers, concluded that the peak temperature gave the best agreement with data from other sources. However, their melting points, determined from the endothermic peak temperature, were often up to 10 °C high. The reason for these divergent views can now generally be explained by the construction of the various pieces of apparatus; in particular, whether the DTA curve is plotted against the sample, reference, or furnace temperature on the abscissa. This problem will be discussed in a following section on interpretation of DTA curves.

Prior to 1955, the normal practice was to immerse the differential thermocouple junctions directly in the sample and reference materials. The serious disadvantages of this procedure have been lucidly pointed out in a classic paper by Boersma.[13] His solution was to embed the junctions in a nickel block containing cavities, into which the sample and reference materials were placed. In this way, he claimed that the importance of diluting the sample with reference material and of careful standardization of sample size and packing density was very much reduced. Nowadays, virtually all commercial differential thermal analysers employ the Boersma principle, in that the samples are contained in crucibles which themselves make contact with the thermocouple junctions.

An excellent and detailed review of the progress of DTA covering the period up to 1962 has been published[14] by Mackenzie and Mitchell.

Although DTA had for some time been used to determine the temperature of second-order phase transitions, the measurement of heat capacities in stable phases was described by David in 1964.[15] A comprehensive theory of such measurements was put forward and the possible effects of experimental parameters considered.

In the same year, Watson and his co-workers[16] published an account of a new technique closely related to DTA, which he called 'differential scanning calorimetry'. The significance of this and its relationship to DTA are discussed later.

More recent reviews of the development and applications of DTA have been published by Duval[17] and by Wunderlich.[18]

THE PRESENT STATE OF DTA

Even in 1962, Mackenzie and Mitchell[14] reported that, to their knowledge, about a thousand papers involving the use of DTA had appeared in scientific journals during the past two and a half years. This explosion of interest in thermal methods generally, and DTA in particular, was due, to a large extent, to differential thermal analysers first becoming generally available on the commercial market. Prior to the 1950s it was generally necessary to construct one's own apparatus. The apparent simplicity of the DTA system unfortunately masks the very high degree of engineering precision necessary to construct a differential thermal analyser capable of yielding quantitative data. Accordingly many workers (the authors among them) often abandoned attempts to use DTA after discovering that the problems in building the apparatus were not primarily electronic, as had been suspected, but rather mechanical. Today, an almost embarrassing choice of instruments is available; details of these are discussed later in this book.

Another result of the growing popularity of thermal methods was the setting up of a group called the 'International Confederation for Thermal Analysis', hereafter referred to as ICTA. The Confederation has since organized a series of international conferences on thermal methods, the first of which was held at Aberdeen, Scotland, in 1965. Subsequent meetings have been at Worcester, Massachusetts, U.S.A. in 1968; then Davos, Switzerland in 1971, followed by Budapest, Hungary in 1974 and Kyoto, Japan in 1977. Each conference has in turn generated further interest in the fields of study discussed, leading to a progressive increase in the numbers of papers submitted. Over the same period, two new journals devoted exclusively to reporting work involving the use of thermal methods have been established. These are the *Journal of Thermal Analysis*, which first appeared in 1969, followed in March 1970 by *Thermochimica Acta*.

Another important contribution by ICTA has been the setting up of committees to standardize the nomenclature used in reporting thermal analysis data and to recommend temperature standards for calibration purposes. The findings of the Nomenclature Committee have now been published[19-21] and are almost universally adhered to in papers currently appearing in the scientific press. Meanwhile, a range of substances undergoing transitions at highly reproducible temperatures has been investigated by scientists drawn from 13 different countries, in order to assess their suitability for use as temperature standards in the calibration of thermal analysis equipment. The data obtained were presented[22] at the Third International Congress on Thermal Analysis and an arrangement was made with the U.S. National Bureau of Standards to market three sets each of five standard substances.

The ICTA Nomenclature Committee have defined[19] DTA as being, 'A technique for recording the difference in temperature between a substance and a reference material against either time or temperature as the two specimens are subjected to identical temperature regimes in an environment heated or cooled at a controlled rate'. Experimental data thus obtained are expressed in the form of a DTA curve; the temperature difference between the test and reference materials (ΔT) is plotted on the ordinate, with endothermic processes shown as a downwards peak. Conversely, an exothermic process will appear as an upwards deflection of the DTA curve. Either time or temperature, (t or T) increasing from left to right, may be plotted on the abscissa. The resulting DTA curve thus normally consists of one or more exothermic and/or endothermic peaks, some of which may be superimposed upon one another. The area under any given peak can be used as a quantitative measure of the amount of heat evolved or absorbed by the physical, or chemical change, which has occurred. Since the heat capacities and thermal conductivities of the test and reference samples are seldom identical, a small temperature differential normally exists between them, when the surrounding furnace is heated or cooled at a uniform rate. Accordingly, the base line of the DTA curve is offset from the abscissa, even though no physical or chemical change is occurring. This effect makes it possible to use DTA to measure heat capacities and to detect the temperature of solid-solid phase transitions.

Consequent upon the general availability of commercial differential thermal analysers, the determination of DTA curves has become very much a routine laboratory procedure, comparable to the measurement of infrared absorption spectra. The apparent simplicity of the technique sometimes obscures the fact that interpretation of DTA curves often demands considerable skill and experience. Even with the thermal decomposition of a simple inorganic compound, numerous problems leading to misinterpretation can arise; a case illustrating some of these difficulties has been described by us.[23]

It is therefore seldom safe to rely on the evidence of DTA alone to elucidate a problem. Techniques most commonly employed in conjunction with DTA are thermogravimetry (TG) and evolved gas analysis (EGA); in the former, changes in weight of a sample subjected to a uniform rate of rise of temperature are recorded. In the latter method, products evolved during DTA are removed by a flowing gas stream and are subsequently collected and analysed, using such techniques as infrared absorption spectroscopy or gas chromatography.

DTA AND DSC

During recent years, considerable confusion has arisen as to the significance of the terms 'differential thermal analysis' and 'differential scanning calorimetry', hereafter referred to as DSC. Despite the impression given by some equipment manufacturers, the terms are not synonymous, although the field of application of the two techniques is virtually identical (see Fig. 1.4).

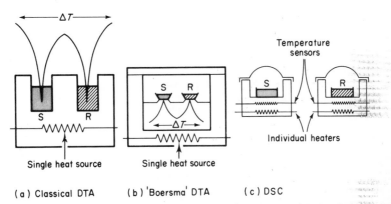

Fig. 1.4. Schematic diagram of the three main differential thermoanalytical techniques. (From Perkin–Elmer *Thermal Analysis Newsletter*, No. 9 of 1970).

Differential scanning calorimetry has been defined by the ICTA Nomenclature Committee[19] as being, 'A technique for recording the energy necessary to establish a zero temperature difference between a substance and a reference material against either time or temperature, as the two specimens are subjected to identical temperature regimes in an environment heated or cooled at a controlled rate'.

It therefore follows that in DSC the sample and reference materials must be equipped with separate heating elements and have separate temperature sensing devices, usually either thermocouples or temperature-sensitive resistors. The two materials are maintained at an identical temperature by electronically controlling the rate at which heat is supplied electrically to the test and reference samples respectively. The ordinate of a DSC curve thus represents the rate of energy absorption by the test sample, relative to that of the reference material; this rate naturally depends on the heat capacity of the sample. The idea of DSC was first put forward by Watson *et al.*[16] in 1964 and has been subsequently exploited by the Perkin-Elmer Corporation, who market commercial DSC apparatus.

For qualitative applications, DTA and DSC are equally good, neither having any obvious advantage in principle. Generally speaking, most commercial differential thermal analysers are capable of covering a much wider temperature range than their DSC counterparts. Conversely, it is claimed that differential scanning calorimeters give better results at very low heating rates.

In quantitative work, wherever measurements can be made by DSC, similar information can be obtained using Boersma type DTA. However, interpretation of the curves obtained differs in some respects, so use of one technique in place of the other should only be undertaken after a full consideration of the theoretical aspects of the problem.

REFERENCES

1. H. Le Châtelier, *Compt. Rend. hebd. Séanc. Acad. Sci. Paris* **104**, 1443 (1887).
2. H. Le Châtelier, *Compt. Rend. hebd. Séanc. Acad. Sci. Paris* **104**, 1517 (1887).
3. H. Le Châtelier, *Bull. Soc. Fr. Miner.* **10**, 204 (1877).
4. H. Le Châtelier, *Z. Phys. Chem.* **1**, 396 (1887).
5. W. C. Roberts-Austen, *Proc. Inst. Mech. Eng.* **1**, 35 (1899).
6. W. C. Roberts-Austen, *Metallographist* **2**, 186 (1899).
7. N. S. Kurnakov, *Z. Anorg. Chem.* **42**, 184 (1904).
8. J. Orcel and S. Caillère, *Compt. Rend. hebd. Séanc. Acad. Sci. Paris* **197**, 744 (1933).
9. L. G. Berg, I. N. Lepeshkov and N. V. Bodaleva, *Dokl. Akad. Nauk. SSSR* **31**, 577 (1941).
10. P. F. Kerr and J. L. Kulp, *Am. Mineral.* **33**, 387 (1948).
11. A. F. Frederickson, *Am. Mineral.* **39**, 1023 (1954).
12. H. Morita and H. M. Rice, *Anal. Chem.* **27**, 336 (1955).
13. S. L. Boersma, *J. Am. Ceram. Soc.* **38**, 281 (1955).
14. R. C. Mackenzie and B. D. Mitchell, *Analyst*, **87**, 420 (1962).
15. D. J. David, *Anal. Chem.* **36**, 2162 (1964).
16. E. S. Watson, M. J. O'Neill, J. Justin and N. Brenner, *Anal. Chem.* **36**, 1233 (1964).
17. C. Duval, *Chim. Anal.* (*Paris*) **54**, 132 (1972).
18. B. Wunderlich, *Thermochim. Acta* **5**, 369 (1973).
19. R. C. Mackenzie, *Talanta* **16**, 1227 (1969).
20. R. C. Mackenzie, *Talanta* **19**, 1079 (1972).
21. R. C. Mackenzie, *J. Therm. Anal.* **8**, 197 (1975).
22. H. G. McAdie, *Proceedings of the Third International Conference for Thermal Analysis*, Davos (1971), Birkhauser, Basel, Switzerland, 1972, pp. 591.
23. M. D. Judd and M. I. Pope, *J. Inorg. Nucl. Chem.* **33**, 365 (1971).

CHAPTER 2

Apparatus

The modern differential thermal analyser is a sophisticated piece of equipment, utilizing many of the advantages made possible by the advent of solid-state electronics. A 'would-be' purchaser is faced with a bewildering range of analysers coupled with an extensive amount of literature from the manufacturers, each claiming that their system offers advantages over their competitors. To someone entering the field of DTA, such a diversity of equipment with differing sample assemblies, furnace design, data recording etc. is simply confusing. However, in essence, the equipment currently available still contains the basic ingredients of apparatus described by Roberts-Austen in 1899.[1]

Figure 2.1 is a block diagram of a DTA apparatus and consists of the following discrete components:

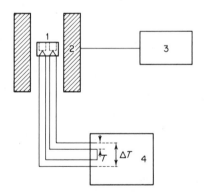

Fig. 2.1. Block diagram of DTA apparatus. 1, Sample holder–measuring system; 2, furnace; 3, temperature programmer; 4, recorder.

1. Sample holder–measuring system. This is the heart of any apparatus. It comprises the thermocouples, sample containers and, in certain cases, a ceramic or metallic block.
2. Furnace—heat source having a large uniform temperature zone.
3. Temperature programmer—to supply energy to the furnace in such a manner as to ensure a reproducible (and preferably linear) rate of change of temperature.

4. Recording system—method of indicating and/or recording the e.m.f.
   (suitably amplified) from the differential and temperature measuring
   thermocouples.

These, therefore, are the essentials for any DTA equipment. The shape, size and performance of each of these is dependent upon many factors, not least among which is the users' requirement. Unfortunately, any alteration in design in one component will be reflected elsewhere and this is particularly true when one considers the sample holder–measuring system. It is therefore necessary to consider, in general terms, what variations are available and what, if any, are the limitations of particular designs.

SAMPLE HOLDER–MEASURING SYSTEM

It is perhaps here that most improvements have taken place. Better methods of amplifying the ΔT signal have meant that increased sensitivity can now be obtained. This, in turn, has led to the use of very small samples, with a consequent reduction in the size of sample holder necessary.

There are, in effect, two quite separate configurations which must be considered and these are effectively controlled by the temperature range under investigation. The requirements for a DTA apparatus covering the temperature range around 1000 °C and upwards are quite different from those when the temperature ranges from -200 °C to $+500$ °C. This will be illustrated below.

Temperatures of 1000 °C or more

Examples of sample holders which are used in this temperature range are shown in Fig. 2.2. In each case it can be seen that the assembly consists of the two thermocouples (one for the sample and one for the reference material) surrounded by some form of block to ensure even heat distribution. Sometimes a third thermocouple is inserted for temperature measurement to avoid using either the sample or reference material temperature. It is doubtful, however, whether this offers any significant improvement in the results obtained.

It will also readily be apparent from Fig. 2.2 that in each case the sample is contained in a small crucible. This crucible is so designed that the base has a small identation in it to ensure a snug fit over the thermocouple bead. Materials used in crucible manufacture include pyrex, silica, nickel and platinum. The choice of crucible is dictated by the temperature range to be investigated and the nature of the sample.

Some more recent equipment incorporates disc-shaped thermocouples with crucibles in the form of a flat low-sided pan. This arrangement is however more commonly utilized for low-temperature DTA (p. 13).

In many early systems the sample was not placed in a container but was in intimate contact with the thermocouples through packing the material into

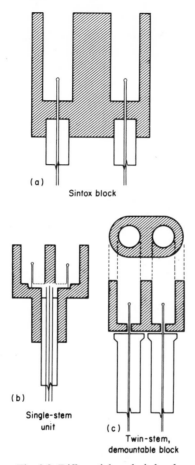

(a)

Sintox block

(b)

Single-stem
unit

(c)

Twin-stem,
demountable block

Fig. 2.2. Differential analysis heads.

recesses in the sample block. Although this practice can lead to an increase in the sensitivity of the system it should be approached with caution since it is unsuitable for quantitative work. Serious contamination problems can occur if the sample melts, or if the degradation products react with either the block material or the thermocouple. If this increased sensitivity is required then it is preferable that the sample should be packed in the crucible and the thermocouples inserted from above. This, unfortunately, may require specialized equipment such as that described in Ref. 2.

The choice of block material is a matter for experiment. Experience has shown that large ceramic blocks (Fig. 2.2(a)) lead to marked reductions in sensitivity and an overall smoothing of the DTA peak. Smaller circular ceramic blocks (Fig. 2.2(b)) with the thermocouples taken up through a single ceramic stem appear to offer the best performance although the authors have found that, with their own equipment, such a block assembly is not the best for quantitative work.

It was, in fact, observed that the most reproducible system for such analyses consisted of a small demountable ceramic block (Fig. 2.2(c)) with the thermo-couples fed through individual stems. However, this only refers to one piece of equipment and cannot be taken to apply to every system. It does show that no one design is ideal for every requirement and that only experience with one's own equipment can give the preferred assembly.

Other materials commonly used for the fabrication of sample blocks are nickel and stainless steel. Metallic blocks have advantages in that they are easily constructed, are non-porous and give rise to little base-line drift.[3] Unfortunately, because of the rapid transfer of heat through the sample and from sample to reference, peaks tend to be smaller than those obtained with comparably sized ceramic blocks.

Thermocouple materials are dictated by the temperature range under study. For temperatures below about 800 °C (or 950 °C in an inert atmosphere) chromel–alumel couples are preferred because of their relatively high e.m.f. (approx. 4 mv per 100 °C). However, most manufacturers these days offer platinum–platinum-rhodium couples as standard items since they have the advantage of a high maximum temperature (up to 1500 °C). The lower output from such thermocouples (approx. 1 mv per 100 °C) is now no longer a handicap owing to the more stable and more sensitive amplifiers currently available.

In the majority of applications, the sample assembly is protected from direct contact with the furnace wall by the addition of a protective ceramic sheath. Such a sheath may be enclosed at one end to provide a static atmosphere or to attain a moderate vacuum, or may be open at both ends for the introduction of a flowing gas atmosphere. The advantages of being able to change the environment around the sample are fully described in the next Section. The use of such a sheath has one other particular advantage. At temperatures above 1000 °C, when heating depends more on radiation than conduction, it is often found that changes in the furnace current give rise to considerable electrical interference. An effective method of preventing this is to coat the sheath over an area surrounding the sample with an earthed platinum film, which can be applied in the form of a paste. Metallic dispersions suitable for this purpose can be obtained from Engelhard Sales Limited, Cinderford, Gloucestershire, England.

Temperature range of −200 °C to +500 °C

The introduction of equipment which operates at subambient temperatures has necessitated a complete redesign of the sample holder–measuring system. A major problem was encountered in transferring heat uniformly away from the sample; this clearly eliminated the use of any heat sink in the form of ceramic or metallic sample block. In general, most manufacturers have adopted designs similar to that shown in Fig. 2.3. The thermocouples (which may be manufactured from either chromel–alumel or iron–constantan) are now usually in the form of a flat disc to ensure the optimum thermal contact with the flat-bottomed sample containers. As before, the sample container may be formed from many different types of material and the choice is dependent on the sample being studied, most common materials are aluminium or platinum foil. It is common practice to enclose the sample by crimping a lid onto the crucible but this is a matter of choice. The much larger area of thermal contact between thermocouple and crucible ensures that very high sensitivity can be attained, with a consequent reduction in the amount of sample used; samples weighing <5 mg are common (although this is not always an advantage).

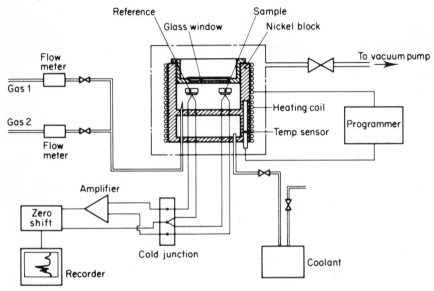

Fig. 2.3. Typical layout of a low-temperature DTA. (By courtesy of Stanton Redcroft, Ltd.)

A major problem with such a system occurs in ensuring that the container is accurately located over the thermocouple and is in the same position from experiment to experiment. This is especially important in the case of quantitative work and caution should be exercised here.

The separate sheath, as described on p. 12, is now no longer required, its place being taken by an enclosed furnace design. This is described below.

THE FURNACE

The main requirements for a furnace for DTA are:

1. A uniform hot zone extending over a wide area of the overall furnace length.
2. Capability of heating up (and in some cases cooling down) at a rate dictated by the temperature programmer, without any appreciable thermal lag.
3. A design such that it can be removed easily from the equipment whilst still hot (<500 °C).

4. It should be so designed that it will be located in the same position relative to the sample holder for successive experiments.
5. It should be wound non-inductively to avoid electrical pick-up by the thermocouples.

As was shown to be the case with the sample holder, furnace design is also affected by the temperature range under investigation.

Temperatures of 1000 °C or more

Nichrome wire is normally used for temperatures not exceeding 1000 °C and platinum wire for use up to 1450 °C. Other combinations have been used, namely Kanthal wire (up to 1350 °C), SiC strip elements (up to 2000 °C) and platinum–rhodium wire (up to 1350 °C) but these are not so common. Better temperature control and more rapid cooling are generally achieved by surrounding the outer casing of the furnace with a water-cooling coil. This latter point is worth considering, since the time taken for a furnace to cool down from high temperature is critical in determining the number of experiments which can be carried out per day.

A major drawback, too, with the larger size of furnace is the thermal lag which is found at low temperatures. This leads to a continually varying heating rate over the temperature range ambient to c. 200 °C. As will be seen later, changes in heating rate can lead to changes in the shape and size of a DTA peak and therefore care has to be exercised wherever the heating rate is not constant.

Many of these problems are eliminated by the use of smaller furnaces such as are found on some commercial equipment. They have the advantages of small size with consequent increases in the possible rates of heating and cooling. Against this must be offset the much smaller hot zone available and it is also doubtful if these smaller furnaces achieve the same temperature stability.

Subambient temperatures

The need to attain temperatures below 20 °C, and to be able to programme down to and up from as low as −200 °C led to a radical change in furnace design. In order to reach these very low temperatures with the capability of programmed heating and cooling it is necessary to design the heating cell to operate both at high temperature (as a normal furnace) and at low temperature using normally liquid nitrogen as a coolant.

A typical furnace design is shown in Fig. 2.3. Liquid nitrogen is allowed in, via a needle valve, to a sheath surrounding, and isolated from, the furnace winding (usually nichrome). Heat is applied to the windings at a steady rate until a balance is reached between heat applied from the windings and heat removed by the liquid nitrogen. Thermal programming is possible by keeping the nitrogen flow-rate constant and varying the heat input. (It will be apparent that some initial practice is required to establish the correct balance.)

To eliminate problems with condensation from the atmosphere onto the sample, the whole furnace assembly is enclosed and sealed around the sample assembly (usually via an 'O' ring seal). This enables different atmospheric conditions to be established without the need for any additional sheath. However, all gases entering the sample environs should be passed through suitable traps to remove any moisture present.

A useful facility included in many low-temperature analysers is a glass top to the furnace. This enables the sample to be observed during heating and cooling; one manufacturer does in fact offer a microscope attachment to their low-temperature equipment.

Because of the very sensitive nature of the thermocouple system used with these small enclosed furnaces, great care must be taken in attaining the required experimental conditions. It is often found that changes in the flow rate of the sample atmosphere and any variation in the heating–cooling balance of the furnace will markedly affect the resultant DTA trace. It would be fair to say that many of the theoretical advantages of this type of system are lost through experimental difficulties.

THE TEMPERATURE PROGRAMMER

The basic requirements for a temperature programmer may be summarized as follows:

1. A wide range of programme rates, e.g. 1–20 °C min^{-1}—easily changed during an experiment.
2. It should be capable of programmed heating and cooling—programmed cooling below 500 °C is not always possible, particularly if large furnaces are used.
3. It should be capable of isothermal control.
4. Proportional control is desirable to eliminate thermal overshoot—a particular problem with large furnaces, as has already been described.
5. The furnace should not generate any electrical interference affecting the ΔT signal.

The ideal temperature programmer will fulfil all these functions and most manufacturers supply these as standard. This is not to imply, however, that equipment cannot be used which does not meet this ideal. For many years, programmers were used (normally cam-driven) where only one heating rate could be selected at a time (others were possible but necessitated some time spent on changing gears or cams). In fact, for the majority of DTA experiments, it may well be found that single heating rate of 10 °C min^{-1} is perfectly adequate. However, the ability to change the heating rate at will is an advantage and frequently yields useful information (see Chaper 3). In fact, when the total cost of a DTA apparatus is considered, a saving of almost certainly less than £200 by including a less refined programmer is hardly worth considering.

RECORDING SYSTEMS

Under this heading we include the different methods which can be used to record the ΔT and T signals together with a brief description of how the signal is amplified prior to recording.

The amplifier

Owing to the very small e.m.f.'s produced by the differential thermocouples it is necessary to amplify the signal by suitable means. In some simple equipment, where large samples (approx. 5 g) are used together with chromel–alumel thermocouples, it is possible to avoid the use of any form of amplification. However, all commercial DTA equipment comes complete with suitable signal amplifiers and the electronic design is now such that very low signal to noise ratios are common for even the most sensitive ranges. In general, the amplifier comes in a form such that several e.m.f. ranges are possible, for instance 25 μV, 50 μV, 100 μV, 250 μV and 500 μV. This simply means that, in the case of the 25 μV range, a signal of 25 μV from the differential thermocouples will produce a full-scale deflection on the recorder supplied with the equipment. Obviously, in the figures quoted above, 25 μV is the most sensitive range and 500 μV the least sensitive. There is no one optimum range setting, but experience will soon show which is the best range for initial experiments on unknown materials.

Potentiometric recorder

As will have already been gathered from the preceding section, the most widely used method of recording the ΔT and T traces is by using a potentiometric recorder. This is the standard method supplied with all commercial equipment. The range of recorders currently available is so wide that any number of recorders will be found adequate. However, from experience, it would appear that the following features should be included in any recorder for DTA.

(a) Twin pen—at first sight this would appear to be an extra expense adding perhaps another £300 or more to the equipment cost. Certainly equipment is available where one pen can be used to record both traces (i.e. the recorder switches from one signal to another at predetermined time intervals). However there is no substitute for having a continuous record of both the ΔT and T traces and any other system must be viewed with caution. It should perhaps be added here that some equipment is supplied with an X–Y recorder where the T signal is recorded on the X-axis and the ΔT signal on the Y axis. It is a matter of personal preference as to whether this form of output is used or whether both T and ΔT are plotted as a function of time.

(b) Adequate sensitivity—most current DTA equipment uses either a 1 mV or a 10 mV recorder. In some cases it is advantageous to be able to extend the sensitivity range without adjusting the amplifier. However this provision is not essential.

(c) Multispeed—for most experiments at heating rates of $10\,°C\,min^{-1}$ a chart speed of $30\,cm\,hr^{-1}$ is satisfactory. However, as has already been referred to above, variations in heating rate can frequently yield useful information. Therefore it is advantageous to be able to compensate for any change in heating rate by a change in chart speed, thus keeping the DTA trace of a reasonable size.

(d) Fast pen response—although the requirement for DTA is much less exacting than for, say, gas chromatography, equipment should be sought with a pen response of better than 1 s for a 10 in chart width. This is especially true when investigating exothermic reactions where occasionally heat is liberated with extreme rapidity.

Indicating galvanometer

This is by far the most simple and least expensive method of obtaining the ΔT signal. A centre zero galvanometer is connected across the ΔT junctions and an operator records the galvanometer deflection as a function of temperature. Although the equipment cost is low, the labour cost is high and thus this method is rarely used in industrial or research laboratories. In teaching laboratories where labour costs are not involved and a small budget does not run to a twin-pen recorder, this method could prove of interest.

Computerized data processing

Such a system is usually in addition to the normal recorded DTA trace and is frequently used for the direct on-line calculation of peak areas, sample purity, etc. Although very useful where numerous experiments of the same form are carried out in a routine operation, it is unlikely that for the majority of users such extra expense could be justified.

SPECIALIZED EQUIPMENT FOR USING DTA UNDER HIGH PRESSURE AND HIGH HEATING RATE

High-pressure DTA

The Clapeyron Equation relates the temperature of a transition to the pressure around the material:

$$\frac{dT}{dP} = \frac{T(V_1 - V_2)}{\Delta H}$$

where T = absolute temperature, P = pressure, $V_1 - V_2$ = volume change in the material as it transforms from phase 1 to phase 2, and ΔH = Change in enthalpy associated with the phase transition.

This equation predicts that if there is a volume change associated with a phase transition, then the temperature at which the transition occurs will vary with the applied pressure. Obviously phase changes such as boiling and sublimation, wheıe $V_1 - V_2$ is large, will be greatly affected whereas solid–solid transitions will be influenced to a far lesser extent.

The earliest commercial equipment to offer a high-pressure facility was the Stone differential thermal analyser. A schematic diagram of the equipment is given in Fig. 2.4. The gas supply is connected to the holder by tubing through a bleed type valve. This valve allows gas to flow to the holder, or to bleed the gas to the atmosphere. The back-pressure regulator is used to control the pressure in the sample holder. This sample holder was designed to be operated at pressures of up to 3000 p.s.i.g. with an upper limit on temperature of 500 °C.

High-pressure DTA has been used to study phase transitions,[4] propellants[5] and effects on the dehydration of such materials as $Mg_2SO_4 \cdot 7H_2O$.[6] In the last case it was shown clearly that the mechanism of dehydration was changed by an increase in pressure.

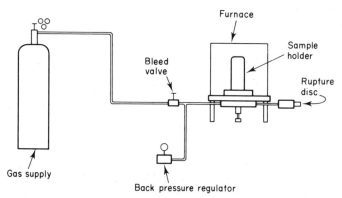

Fig. 2.4. Schematic diagram of high-pressure sample holder set-up. (Courtesy of Columbia Scientific Industries, Texas, U.S.A.)

Equipment for DTA under pressure is now widely available and offers many interesting fields for investigation. Not least of these are the explosives, which can be handled safely in equipment built to the rugged standards required by high-pressure DTA.

High heating rate DTA

In general, heating rates of the order of 10 °C min⁻¹ are used for conventional DTA studies. Changing the heating rate and its effect on a DTA trace is described in Chapter 3 but in general this effect is rarely investigated. To a large extent, the choice of heating rate is determined by the equipment that is available, and is generally fixed by the furnace design. Even with the new small mass furnaces it is unusual for heating rates of >20 °C min⁻¹ to be available, owing to problems with base-line drift as the sample temperature changes relative to the reference material.

However, the use of high heating/cooling rates would offer distinct advantages as far as time saving is concerned. A very elegant piece of equipment offering this facility has been described by Miller and Sommer[7] and is incorporated into a hot-stage microscope.

The equipment consists of a modification of the hot-stage microscope devised by Welch[8] in which a small thermocouple acts both as a furnace and temperature sensor. Earlier work by Mercer and Miller[9] had adapted Welch's design to incorporate continuous monitoring and recording of the heating and cooling curves for microscopic quantities of materials, supported in the thermocouple microfurnace. The rate of change of temperature could be varied over a wide range with an air-quenching rate of 1000 °C min^{-1} being achieved.

The equipment is based on the arrangement used by Welch whereby the thermocouple e.m.f. could be isolated from the heating current. The furnace assembly consists of two thermocouples functioning as microfurnaces, thermometers and supports for sample and reference material, mounted in a windowed cell on a microscope stage. The e.m.f. from the sample thermocouple, and the differential e.m.f. are displayed on a recorder with very high chart speed (1 in s^{-1}). The basic circuit diagram is shown in Fig. 2.5.

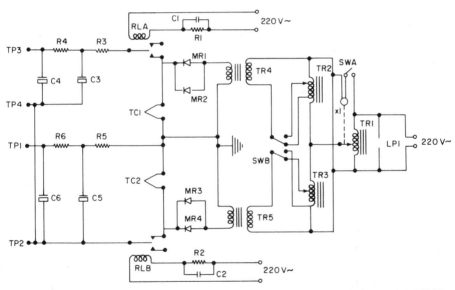

Fig. 2.5. Circuit Diagram. SWA, switch S.P.S.T. 220 V, 2.5 A; SWB, switch D.P.S.T. 220 V, 2.5 A; UP1, pilot lamp, 220 V; TR1, autotransformer, motor-driven, 220 V, 2.5 A; TR2, autotransformer, dual-track, 0–220 V, 2.5 A (Zenith V511/2B); TR3, autotransformer, dual-track, 0–220 V, 2.5 A (Zenith V511/2B); TR4, TR5, transformers; 220 V primary, 12 V, 5 A secondary; MR1, MR2, MR3, MR4, rectifier, silicon, 2.5 A (Semikron SK 2.5/10 or Texas Instrument IS 401); C1, C2, capacitor, fixed, 250 V, 0.26 μF; C3, C4, C5, C6, capacitor, fixed, electrolytic, 12 V, 2000 ohm; R1, R2, resistor, fixed, carbon, 10 W, 4000 ohm; R3, R4, R5, R6, resistor, fixed, carbon, 0.5 W, 10 ohm; RLA, RLB, synchronous converter; TC1, TC2, thermocouples; TP1, TP2, output connectors for measurement of sample thermoelectric e.m.f.; TPC, TP4, output connectors for measurement of differential thermoelectric e.m.f.; X1, motor for TR1. (From Miller and Sommer, Ref. 7.)

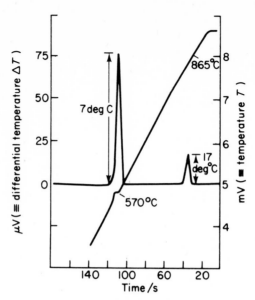

Fig. 2.6. Thermal analysis of lithium sulphate compared with borax. Thermocouples, Pt/10% Rh-Pt; chart speed, $1\frac{1}{2}$ in min^{-1}; rate of cool between 865 and 570°C, 3.7°C s^{-1}; weight of Li$_2$SO$_4$, 200 μg. (Miller and Sommer, Ref. 7.)

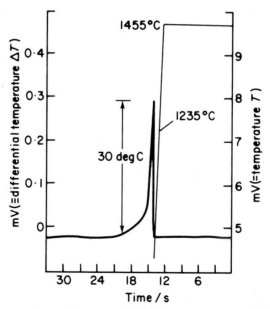

Fig. 2.7. Thermal analysis during quenching of eutectic melt 2CaO: 3BeO (empty reference junction). Thermocouples, 6% Rh Pt/30 Rh-Pt: chart speed 200 mm min^{-1}; rate of cool, 400°C s^{-1}; weight 0 (2CaO: 3BeO)~20 μg. (Miller and Sommer, Ref. 7.)

Figures 2.6 and 2.7 serve to demonstrate the efficiency of such a system. The sample weights involved are very small and the cooling rates very fast, yet the resolution is excellent. It is only unfortunate that this equipment is not available commercially, although the cells described above can be obtained from Tele-communication Instruments Limited, Great Yarmouth, U.K.

REFERENCES

1. W. C. Roberts-Austen, *Proc. Inst. Mech. Eng. (London)* **1**, 35 (1899).
2. B. W. Neate, D. Elwell, S. H. Smith and M. D'Agostino, *J. Phys. E.* **4**, 775 (1971).
3. A. F. Frederickson, *Am. Mineral.* **39**, 1023 (1954).
4. E. Rapoport and G. C. Kennedy, *J. Phys. Chem. Solids* **27**, 93 (1966).
5. R. H. Bohon, *Anal. Chem.* **35**, 1845 (1963).
6. C. E. Locke and R. L. Stone, *Thermal Analysis* Vol. II (*Proceedings of the Second International Conference for Thermal Analysis*). Academic Press, New York, 1969, p. 963.
7. R. P. Miller and G. Sommer, *J. Sci. Instrum.* **43**, 293 (1966).
8. J. H. Welch, *J. Sci. Instrum.* **31**, 458 (1954).
9. R. A. Mercer and R. P. Miller, *J. Sci. Instrum.* **40**, 352 (1963).

CHAPTER 3

Experimental Factors

The effects of variations in 'experimental' parameters on the shape of a DTA trace are numerous and it is not practicable to describe all of them in a chapter of this size. This chapter is therefore restricted to the more 'practical' of these variables, namely effects associated with the sample itself, with the atmosphere surrounding the sample and, to a lesser extent, with the rate of heat supply to the sample. Other factors are known to affect the DTA trace, but these can be attributed to equipment variables, such as sample holder design, furnace design, etc. It is assumed that the reader has, or will obtain, custom built equipment and therefore has, to some extent, these equipment variables fixed by his choice of apparatus. Such variables therefore will not be covered and this chapter is restricted to those variables over which the experimenter has direct control.

Let us first consider an example which may be used to demonstrate some of the above-mentioned effects. The decomposition of zinc oxalate dihydrate has been studied by Dollimore et al.,[1,2] who showed that, in an atmosphere of either nitrogen or oxygen, zinc oxide is formed as the end product. This reaction was shown to be endothermic in nitrogen and exothermic in oxygen; the exothermic reaction was attributed to the oxidation of the carbon monoxide, formed during the decomposition, to carbon dioxide at the surface of the zinc oxide.

Figure 3.1 gives the DTA traces obtained[3] for the decomposition of zinc oxalate dihydrate under various experimental conditions. (The dehydration peak is endothermic in all cases and this section of the curves has been omitted.) These traces clearly demonstrate the effects which can be obtained by varying the sample weight, sample atmosphere and heating rate.

1. *Variation in sample atmosphere* (sample weight and heating rate constant)— Fig. 3.1 (c) and (d). The traces indicate that the reaction is exothermic in an atmosphere of oxygen and endothermic where the atmosphere is nitrogen. This exothermic reaction is attributed to the oxidation of carbon monoxide, as mentioned above. A consideration of the respective peak areas shows that the heat changes are in the approximate ratio of 2.5:1 in favour of the oxidation process and this, therefore, would be expected to predominate in any DTA trace obtained by decomposition of zinc oxalate in air.

2. *Variation of heating rate* (sample weight and sample atmosphere constant)— Fig. 3.1 (b) and (e). At a heating rate of 10 °C min⁻¹ and in an atmosphere of air,

DTA curves

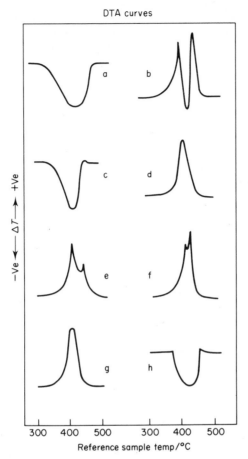

Fig. 3.1. Differential thermal analysis of zinc oxalate dihydrate under various experimental conditions. (The dehydration peak is endothermic in all cases and this section of the curves has been omitted). (a) Sample weight 1.00 g, heating rate 5°C min^{-1}, static air. (b) Sample weight 95 mg, heating rate 10°C min^{-1}, static air. (c) Sample weight 82 mg, heating rate 10°C min^{-1}, nitrogen (1000 cm^3 min^{-1}). (d) Sample weight 82 mg, heating rate 10°C min^{-1}, oxygen (1000 cm^3/min). (e) Sample weight 97 mg, heating rate 5°C min^{-1}, static air. (f) Sample weight 80 mg, heating rate 5°C min^{-1}, static air. (g) Sample weight 66 mg, heating rate 5°C min^{-1}, static air. (h) Sample weight 206 mg, heating rate 5°C min^{-1}, static air. (Judd and Pope, Ref. 3.)

the trace is initially exothermic and the endothermic contribution is only apparent when the decomposition process is well under way. A stage is then reached when the rate of evolution of heat by the oxidation reaction is exceeded by the rate of absorption of heat by the endothermic decomposition process. When the heating rate is reduced, the rate of evolution of gaseous products also falls and the endothermic decomposition process has a much smaller effect on the DTA curve.

3. *Variation in sample size* (heating rate and sample atmosphere constant)—Fig. 3.1 (e), (f), (g) and (h). A factor which clearly affects the reaction is the rate

of diffusion of either oxygen to the sample, or of gaseous decomposition products into the atmosphere. For the exothermic reaction to be observed, oxygen and carbon monoxide must react in the sample container so that the heat evolved can be measured by the thermocouple. Decreasing the sample weight facilitates this process since the oxygen then has an easier access to the sample. Conversely, an increased sample weight increases the endothermic contribution, as shown in Fig. 3.1 (h).

This example, therefore, demonstrates the care that is necessary to ensure that the experimental conditions chosen for the investigation are realistic. It should be stated here that the effects of these variables are most noticeable in the case of decomposition reactions; where phase transitions are involved the effects are far less marked or even non-existent.

FACTORS ASSOCIATED WITH THE SAMPLE

From the example given it has been shown that, for a decomposition reaction, the amount of sample used can have a marked effect on the result obtained. However, it is not just sample quantity that is involved; the physical nature of the sample and the way that is packed into the sample container are very important variables. To a lesser extent, the pretreatment of the sample prior to the analysis is also important, especially where materials which hydrate easily are involved.

Differential thermal analysis can be used to study any material which can be produced in such a form as to enable it to be placed in a sample crucible. However, the vast majority of experiments are performed on materials in powder form and therefore the first variable to be considered is the effect of particle size. Unfortunately it is not possible to be specific as to what effect particle size has on the shape of a DTA trace, since this is dependent on the type of chemical process under investigation. For surface reactions and those reactions which are diffusion controlled, changes in particle size may lead to changes in peak temperature, with decreasing peak temperature as the particle size is reduced. In the case of decomposition reactions which proceed via a mechanism other than diffusion control, and for phase transitions, the effect of particle size is generally minimal.

This is well illustrated by the decomposition of $CuSO_4.5H_2O$ to $CuSO_4.H_2O$, using samples of constant weight, with a heating rate of 10 °C min^{-1} and a static nitrogen atmosphere. The curves shown in Fig. 3.2 were obtained[4] using a Standata 6-25 DTA apparatus. Three separate endothermic processes contribute to the observed endothermic peaks:[10]

(1) $CuSO_4.5H_2O \rightarrow CuSO_4.3H_2O + 2H_2O$ (liquid)
(2) H_2O (liquid) $\rightarrow H_2O$ (gas)
(3) $CuSO_4.3H_2O \rightarrow CuSO_4.H_2O + 2H_2O$ (gas)

With the largest particles ($-14+18$ BS mesh), the first stage of decomposition occurs comparatively slowly, because of the time taken for the evolved water to diffuse to the surface. The corresponding endothermic peak then tends to merge with that due to boiling of the water, the peak temperature of which was found to be independent of particle size (Fig. 3.2(a)). This is followed by the endotherm due to reaction (3).

Samples having a rather smaller particle size ($-52+72$ BS mesh) showed three distinct and separate endothermic peaks (Fig. 3.2(b)). Reaction (1) now occurs more rapidly and is virtually complete by the temperature at which the endotherm due to boiling becomes apparent. The peak temperature corresponding to reaction (3) is also slightly reduced.

Using even smaller particles ($-72+100$ BS mesh), only two endotherms could be detected (Fig. 3.2(c)). Reaction (1) now occurs at a lower temperature

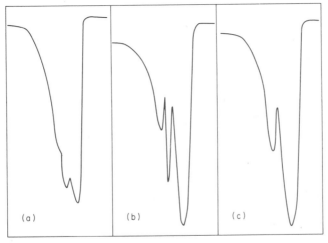

Fig. 3.2. Differential thermal analysis curves in nitrogen for the reaction $CuSO_4.5H_2O \rightarrow CuSO_4.H_2O + 4H_2O$ heated at $10°C$ min^{-1}, for (a) $-14+18$, (b) $-52+72$, and (c) $-72+100$ BS sieve fractions.

still, as also does reaction (3). This results in the boiling endotherm (2) becoming completely obscured by that due to step (3), giving the appearance of a two-stage process.

When using samples with a wide particle size distribution, under the same experimental conditions, three distinct peaks were always observed. Reducing the heating rate tends to minimize the effects illustrated in Fig. 3.2, because the reactions occur more slowly and are therefore less subject to inhibition by the rate of diffusion of volatile products to the surface of the particles.

Clearly, then, the shape of a DTA curve can show marked variations with particle size, even when the sample mass, ambient atmosphere and heating rate are held constant. It is not possible to be dogmatic as to the correct particle size range to select for DTA, although finer powders are generally preferred. What is important is to ensure that, when comparing similar materials, each sample consists of particles within the same size range.

Having discussed the particle size of the sample, the next variable to be considered is the method of packing the sample into the container. Curves for materials undergoing reaction with the atmosphere surrounding the sample are strongly affected by the packing of the sample, since access of the atmosphere to the reacting material is impeded. It is therefore important that a reproducible method of sample packing be utilized. Of the many methods which are available, the authors' preference is for hand tamping, where the loaded sample container is tapped on the bench for a fixed number of times. At first sight this may sound a faintly ridiculous and time-consuming method, but it very quickly becomes a part of the experimental routine.

The effect of sample weight (or bulk) has already been described for one particular process. In general, large samples lead to an increase in sensitivity (owing to the higher heat change) but with a corresponding decrease in resolution. In the early days of DTA large samples had to be employed, since the instrumentation available could only measure relatively large thermocouple e.m.f.'s. However, nowadays the small signals obtained by the use of smaller samples do not pose any instrumentation problems. The advantages offered by small samples include sharper peaks and more resolution between the peaks, coupled to a distinct reduction in base-line drift (as the properties of the sample approach more closely those of the reference material). In cases where precious metal compounds are being studied, the advantages are not just practical but also pecuniary!

Where it is not practical to use a smaller sample of material because of instrumental problems (only large sample containers available, possibly), then the sample may be diluted with reference material. This can be achieved either by the preparation of an intimate mixture, or by the process of layering,[5] where the crucible is filled with alternate layers of reference and sample material. Dilution is particularly important where violent reactions are involved (such as explosive decomposition), or where the sample shrinks away from the wall of the sample container. However, when a diluent is used, it is necessary to ensure that the material is truly inert and will not in any way interact with the reaction products. The adsorption of gases produced during a reaction by a diluent has been observed[6] by some workers.

Where materials are subject to hydration by atmospheric water vapour, such as hydrates, some clays and certain plastics such as nylon, it may well prove necessary to pretreat the sample under a set of standard conditions prior to analysis. For clays, MacKenzie[7] recommends equilibrating the sample over a saturated solution of $Mg(NO_3)_2.6H_2O$. This gives a relative humidity of 55% at 18 °C and 51.5% at 30 °C and is thus little affected by changes in the ambient temperature.

The effects mentioned in the previous paragraphs are particularly apparent with decomposition processes and, in general, have little or no effect on traces obtained for phase transitions (other than increases or decreases in peak height). In fact, any change in peak temperature with sample size can be taken as evidence that that peak is not associated with a phase transition. It is not possible to recommend an ideal sample weight since this is, to some extent, dictated by the equipment which is to be used. The most realistic approach would be to use some initial experiments to decide practically the sample weight which is appropriate for the reaction under investigation. For instance, the authors found that in their work on the decomposition of the alkaline earth carbonate, a sample weight in the range 60–80 mg was most satisfactory.

FACTORS ASSOCIATED WITH THE ATMOSPHERE SURROUNDING THE SAMPLE

The atmosphere around the sample may be either static, or flowing and may be either inert, or reactive. In general the atmosphere is chosen to be compatible with the process under investigation. In other words, where it is known that no reaction of the sample degradation products occurs in air, then it is logical to use air as the sample atmosphere. An example of this type of process would be in studying phase transitions where the components are stable in air at high temperature. Where a decomposition process is to be investigated and especially where a compound containing carbon is involved, then it is advisable to use an inert atmosphere, such as nitrogen, in order to avoid any oxidation of residual carbon, or carbon monoxide (as shown in the example given earlier).

Most workers adopt the principle of flowing an inert gas over the sample, allowing adequate time for flushing the air out of the system prior to commencement of the experiment. The ideal situation would be where the gas flows through the sample as well as around it and in this context sample buckets made of a fine wire mesh have been described.[8] Flow rates should be such that the thermal conditions around the sample are not altered; in most cases manufacturers give recommended values for their equipment.

One atmosphere which has received little attention in DTA is vacuum. The use of a vacuum has many advantages, not least of which is that decomposition reactions can be made to occur at much lower temperatures than they normally occur in air. For instance, the decomposition of calcium carbonate has been shown[9] to commence as early as 550 °C under vacuum (as compared to 700 °C in air) due to the rapid removal of CO_2 from above the sample. A further advantage is that evolved gas detection (EGD) can be performed easily by installing a suitable vacuum gauge into the pumping line. This is demonstrated in Fig. 3.3 for the decomposition of calcium carbonate; in this case the evolved gas detector was a Pirani gauge.

Unfortunately vacuum DTA is beset with problems, most of which can be attributed to difficulties in heat transfer through a vacuum to the sample assembly. In the authors' experience with their own equipment, under a pressure of

10^{-3} Torr it was not uncommon to find a thermal lag of up to 50 °C between the sample and the furnace, although this decreased to less than 10 °C at temperatures above 300 °C. Further problems occur with ejection of the sample from its container, movement of the container over the thermocouple giving a

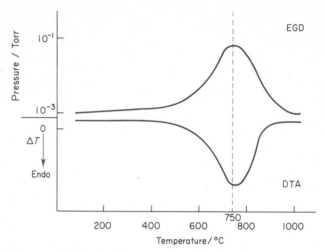

Fig. 3.3. Simultaneous DTA/EGD for a sample of CaCO₃ (75 mg) heated at 10°C min⁻¹ under vacuum (initially < 10⁻³ Torr).

ragged base line, and, most difficult of all, the problems involved in obtaining a vacuum-tight system. However, this is an area of DTA still to be explored and it is certain that, in the near future, satisfactory apparatus will become widely available.

FACTORS ASSOCIATED WITH HEATING RATE

Of all the factors which can affect the shape of a DTA trace, this has received least attention, probably due to the fact that a lower heating rate involves more time per experiment.

In general, the effect of lowering the heating rate is the same as that obtained by lowering the sample weight, in that sharper peaks are produced and the resolution is increased. For most applications a heating rate of 10 °C min⁻¹ is adopted, since this appears to offer the best compromise between quality of resolution and time taken per experiment. However, in cases where an accurate temperature measurement is required, such as in the determination of phase transitions, then it may well prove necessary for the experimenter to investigate the effect of heating (or cooling) rate on the DTA trace. Heating rates as low as 2 °C min⁻¹ are not uncommon in this type of study.

As has already been mentioned, in the case of a decomposition process where two or more concurrent reactions are taking place, changing the heating rate may well lead to additional information. If the kinetics of a reaction are to be investigated, it is particularly important to minimize the temperature differential across the sample; this can be achieved either by decreasing the sample weight, or by lowering the heating rate. It is not uncommon to find that a change in the sample mass also affects the observed kinetics. Therefore it is generally preferable to reduce the heating rate instead.

In conclusion, then, it is clear that care has to be taken to ensure that the experimental conditions are well defined before drawing any conclusions from a DTA trace. It is especially important that full details of the sample are known with respect to history, particle size and so on. To obtain a DTA trace is simplicity itself. To obtain meaningful results takes care.

REFERENCES

1. D. Dollimore, D. L. Griffiths and D. Nicholson, *J. Chem. Soc.* 2617 (1963).
2. D. Dollimore and D. Nicholson, *J. Chem. Soc.* 908 (1964).
3. M. D. Judd and M. I. Pope, *J. Inorg. Nucl. Chem.* **33**, 365 (1971).
4. M. I. Pope and D. I. Sutton, *Thermochim. Acta.* in press.
5. O. Talibudeen, *J. Soil Sci.* **3**, 251 (1952).
6. E. M. Barrall and L. B. Rogers, *Anal. Chem.* **34**, 1106 (1962).
7. R. C. MacKenzie and B. D. Mitchell, *Differential Thermal Analysis*, Vol. I, (Ed. R. C. MacKenzie) Academic Press, New York, 1970, p. 113.
8. M. D. Karkhanavala, V. V. Deshpande and A. B. Phadris, *J. Therm. Anal.* **2**, 259 (1970).
9. M. D. Judd and M. I. Pope, *Thermal Analysis* (Proceedings of the Second International Conference of Thermal Analysis) (Eds. R. F. Schwenker and P. D. Garn), 1969, p. 1423.
10. W. W. Wendlandt, *Thermochim. Acta* **1**, 419 (1970).

CHAPTER 4

Interpretation and Presentation of Data

THE DTA CURVE

Part of a simple and idealized DTA curve is illustrated in Fig. 4.1. On raising the temperature of the furnace, a small but steady temperature difference develops between the test and reference materials. This is because, although the temperature at the centres of both materials will lag behind that of the furnace, the magnitude of the lag depends primarily on the thermal conductivity and heat capacity of each substance. Accordingly, the DTA curve continues in an approximately rectilinear manner, until the test material undergoes some physical or chemical change (AB). At B, the curve begins to deviate from the base line, due to an exothermic process occurring within the test sample. This point, B, will hereafter be referred to as the onset temperature of the reaction, or phase transition, since it represents the temperature at which the process first becomes detectable by DTA. The exothermic peak temperature, C, corresponds to the maximum rate of heat evolution detected by the differential thermocouples. It does not necessarily correspond to the maximum rate of reaction, nor to completion of the exothermic process. Generally speaking, the peak temperature

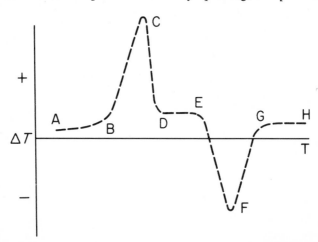

Fig. 4.1. A hypothetical DTA trace showing an exothermic peak, an endothermtic peak and the change in off-set temperature of the baseline resulting from these processes.

30

tends to show a much greater variation with heating rate and other experimental factors than does the onset temperature, although peak temperatures are more easily measured.

The exothermic process giving rise to the peak BCD becomes complete at some temperature between the points C and D. No further evolution of heat is detectable above D, so the curve returns to a new base line, DE. The heights of AB and DE above the abscissa normally differ, reflecting the change in heat capacity of the sample, resulting from the exothermic process which has occurred.

The onset of an endothermic process is indicated by the downwards deflection of the base line at E, giving rise to the endothermic peak EFG. Completion of this process and the formation of a new, thermally stable phase, is confirmed by the horizontal portion of the curve, GH.

In our hypothetical example, Fig. 4.1, the DTA peaks are reasonably sharp and well separated. This is frequently not the case in practice; superimposed and overlapping peaks often occur, making interpretation difficult. Various experimental procedures have been developed to aid peak separation and to increase the sharpness of peaks e.g. by changes in heating rate, sample dilution, increase or decrease in the pressure of the ambient atmosphere, etc. However, interpretation of DTA curves without evidence obtained from other techniques is usually very unwise, except in the case of the simplest systems. This is well illustrated by the variety of curves which can be produced by the thermal decomposition of zinc oxalate dihydrate,[1] itself a far from complex material (see Fig. 3.1).

It is essential to remember that peaks on a DTA curve can arise from both physical and chemical changes. The former include melting, boiling and solid–solid structural transitions; sometimes a solid reaction product will form a low melting eutectic with a sample which would not itself have melted. In the latter case, peaks can result not only from reactions of the sample itself, but also from secondary reactions of evolved decomposition products. Such reactions can often occur catalytically on the surface of the residual solid.

INTERPRETATION OF A SINGLE PEAK

The significance of the various points on a DTA peak depends very much on the position of the thermocouple used to measure the temperature plotted on the abscissa.

It is also essential to know if the differential thermocouples are actually immersed in the sample and reference materials, as was generally the case in classical DTA. Modern apparatus usually adopts the principles set out by Boersma,[2] in which the differential thermocouples are in intimate contact with vessels containing the test and reference materials, respectively.

The effect of thermocouple position on peak shape seems first to have been

considered in detail by Smyth,[3] in 1951. He calculated theoretical peak shapes,
assuming that the temperature measuring thermocouple was located (a) at the
centre of the reference sample, (b) at the surface of the test sample and (c) at the
centre of the test sample. These curves led him to conclude that in case (a),
neither the onset, nor the peak temperature, gave a true measure of the actual
transition temperature. In case (b) the true transition temperature was correctly
obtained from the onset temperature; while in case (c) best agreement was obtained with the peak temperature. However, controversy continued; Frederickson[4] reported in 1954 that the onset temperature gave the best measure of
transition, or reaction temperature, and that this could most accurately be obtained from the first derivative of a DTA curve (Fig. 4.2). Conversely, Morita
and Rice[5] concluded that peak temperatures were most suitable for determining
melting and boiling points of organic materials.

Smyth's predictions were investigated experimentally by Barrall and Rogers[6]

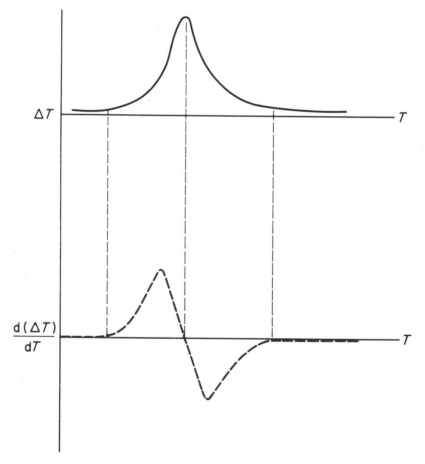

Fig. 4.2. Hypothetical DTA curve and the first derivative of that curve.

in 1962, using the thermocouples embedded in (1) the DTA head, (2) the reference sample and (3) the test sample, for temperature measurement. Their conclusions can be summed up as follows. In case (1) both the sample and reference temperatures lag behind that recorded; if the sample melts, then its temperature lags

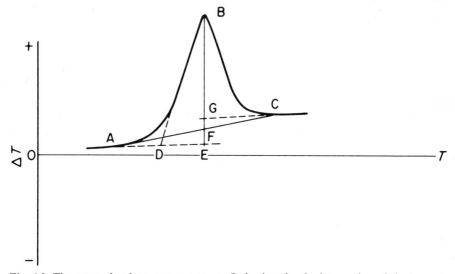

Fig. 4.3. The extrapolated onset temperature, D, is given by the intersection of the tangent drawn at the point of greatest slope on the leading edge of the peak, AB, with the extrapolated baseline AD.

behind that of the reference material. In case (2) the sample and reference temperatures remain closely similar until the sample melts. High heating rates shift the melting endotherm peak towards higher temperatures. Finally, in case (3), the peak temperature is not affected by changes in heating rate, but it does not give a true measure of the sample melting point.

More recent work by Barrall[7] (1973) suggests that melting points obtained from the endotherm onset temperature are usually accurate to within one or two percent, if the sample temperature is plotted on the abscissa. Use of very small samples minimizes thermal gradients. The peak temperature was found to coincide with the disappearance of the last trace of solid, where compounds did not have a sharp melting point.

The rather confusing data previously described all relate to systems where the differential thermocouples are immersed in the test and reference materials. A procedure widely adopted with modern DTA apparatus is illustrated in Fig. 4.3. This is to determine the extrapolated onset temperature, D, which is[8] the point of intersection of the tangent drawn at the point of greatest slope on the leading edge of the peak (AB) with the extrapolated base line (AD). The authors have obtained good agreement between experimental and literature values of melting points using the extrapolated onset temperature; the reference thermocouple temperature was plotted on the abscissa.

The peak area

The peak area is defined[8] by ICTA as the area enclosed between the peak and the interpolated base line, ABCA in Fig. 4.3. However, some workers believe that the area enclosed by a line joining the points ABCGEA gives a more accurate measure of the associated heat change. In most cases, the difference between these two areas is experimentally insignificant.

That the area under a DTA peak is directly proportional to the corresponding enthalpy change appears to have been suggested first by Spiel in 1944. A theory relating peak area to various experimental parameters was subsequently developed by Kerr and Kulp[9] (1948). They confirmed that peak area was indeed a measure of the total enthalpy change and that it was not affected by the heat capacity of the sample. The theory of quantitative DTA was further developed by Borchardt and Daniels,[10] who also showed that the technique could be used to study the kinetics of reactions.

Nevertheless, the use of DTA to obtain calorimetric measurements effectively dates from the classic paper by Boersma,[2] pointing out the consequences of having the differential thermocouple junctions actually immersed in the reactant sample and reference material. For a DTA apparatus where the differential thermocouples are in thermal, but not physical, contact with the test and reference materials, he developed the equation:

$$\text{Peak area} = \int_{T_1}^{T_2} \theta \, dT = \frac{mq}{G}$$

Where m is the sample mass, q the enthalpy change per unit mass and G a heat transfer coefficient; G cannot be calculated theoretically but is obtained by calibration using standard substances.

Since then, numerous papers have appeared on the correlation between peak area and enthalpy change. Recent examples are those of Sturm[11] and of Vaidya and Nicholson;[12] the latter workers in effect equate Boersma's symbol G with gK_s, where g is a shape factor and K_s the thermal conductivity of the sample.

A significant contribution to the interpretation of peak areas has, however, been made by Berg, Egunov and Kiyaev.[13] They have shown that the gas filling the pores and interstices between crystallites can have a pronounced effect on measured enthalpy changes. This is due to a change in the thermal conductivity of the gaseous atmosphere in the vicinity of the DTA sample container. Where a DTA peak is due to a reaction in which gaseous decomposition products are evolved, then the ambient atmosphere may have a significantly different thermal conductivity from that used during calorimetric calibration of the apparatus. Hence serious errors can be introduced. For example, a 25% difference was observed in the peak areas for a given process measured in atmospheres of carbon dioxide and of helium. Their equation relating peak area to thermal conductivity of the gaseous atmosphere is given below.

$$\frac{\text{Peak area in gas, } G_1}{\text{Peak area in gas, } G_2} = \frac{\lambda_S + \lambda_{G1}\left(\dfrac{\gamma_S}{\gamma} - 1\right)}{\lambda_S + \lambda_{G2}\left(\dfrac{\gamma_S}{\gamma} - 1\right)}$$

λ_G and λ_S are the thermal conductivities of the gas G and the massive solid, S, respectively. γ_S is the crystal density of the solid S and γ is the bulk density of the sample.

In certain cases, where the DTA peak is very sharp, narrow and symmetrical, it may be permissible to take the peak height, rather than the peak area, as a measure of the total enthalpy change.

MEASUREMENT OF PEAK AREA

The determination of enthalpy changes by DTA involves precise measurement of the area under the corresponding peak. The accuracy to within which this area can be measured therefore limits the overall accuracy of calorimetric determinations.

There are three basic approaches to the problem of peak area measurement.
1. The 'cut and weigh' method.
2. Use of a planimeter.
3. Use of a pen recorder fitted with a mechanical or electronic integrator.

All these methods are widely used and none, except possibly some recent electronic integrators, has been shown to be capable of consistently greater accuracy, provided adequate precautions are observed.

The 'cut and weigh' method

This involves cutting out the peak recorded on the DTA trace and then weighing the piece of paper under standard conditions. The area is obtained by comparison of the weight with that of a square, of accurately known area, cut from the same roll of recorder paper. Since it is usually desirable to preserve the DTA trace for future reference, the same 'cut and weigh' procedure is more often carried out on a photocopy of the original curve.

The errors introduced by this method of peak area measurement arise mainly through (a) inaccurate cutting out, (b) inhomogeneity in the paper, and (c) variation in the water content of the paper. Probably the last-mentioned factor is the most common source of error, particularly where photocopies are used. All pieces of paper should be stored and weighed under conditions of controlled humidity and temperature.

The planimeter method

This, of course, involves purchasing a planimeter, costing (at 1975 prices) between about £50 and £100. An instrument which has proved very satisfactory for evaluating DTA curves is manufactured by W. F. Stanley and Company Limited, of Avery Hill Road, New Eltham, London, S.E.9. In laboratories where peak area measurement is carried out frequently, the planimeter will soon pay for itself in time saved, compared to the 'cut and weigh' method. To use a planimeter, the operator follows the perimeter of the DTA peak with a stylus, mounted on the planimeter arm. The area enclosed by the peak is then recorded directly on a dial.

Although both speed and reproducibility quickly improve with operator practice, this method is generally capable of an accuracy as good as the best obtainable by the 'cut and weigh' method. Sources of error (b) and (c) are entirely eliminated, without any compensating disadvantages.

Integrating recorders

Some potentiometric recorders, suitable for use with DTA apparatus, can be supplied with integrating devices as an optional extra. Clearly the accuracy obtainable will vary from instrument to instrument and with the integrating method used. It is by no means certain that the mechanical types of integrator generally available are capable of as high a degree of accuracy and reproducibility as the two methods previously described. Electronic integrators, on the other hand, are certainly capable of a higher performance, but at a very substantially increased cost. Whether this can be justified must remain a matter of individual choice.

CALIBRATION FOR CALORIMETRIC MEASUREMENTS

It was shown on p. 34 that a DTA peak area is proportional to the total enthalpy change which has occurred. The proportionality constant cannot yet be calculated theoretically, although many of the factors which influence it are known. The DTA apparatus is therefore calibrated by measuring the peak areas produced by processes which are accompanied by an accurately known enthalpy change. Since the proportionality constant is not independent of temperature, calibration must be carried out using processes which occur within the same temperature range as the experimental measurements. It has been suggested[14, 15] that dilution of the sample with an inert reference material, having a high thermal conductivity, will minimize changes in the proportionality constant with temperature.

Calibration is normally carried out using the enthalpies of fusion of high purity substances, since this avoids any complications which might arise through the behaviour of decomposition products evolved during a chemical reaction. However, melting transitions also have their problems; some substances melt to form a spherical droplet, which makes poor thermal contact with the DTA crucible. In other cases, the melt tends to react with the DTA crucible, which is usually made of platinum. Oxygen must be excluded vigorously from the system to prevent the formation of oxide films on the surface of metals used in calibration.

An example of some substances which may prove useful in the enthalpy calibration of DTA apparatus is given in Table 4.1. These cover the range from

TABLE 4.1

Element or compound	Melting point/°C	Enthalpy of fusion/Jg^{-1}	Data sources Ref.
Naphthalene	80.3	149	20
Benzoic acid	122.4	148	23
Indium	156.6	28.5[a]	14
Tin	231.9	60.7	14
Lead	327.5	22.6	24
Zinc	419.5	113	24
Aluminium	660.2	396	24
Silver	960.8	105	24
Gold	1063.0	62.8	24

[a]This figure has been queried by Richardson and Savill,[25] who claim that a more accurate value is 29.2 Jg^{-1}.

room temperature to about 1100 °C, but care needs to be exercised in the use of lead and aluminium. The former is particularly likely to form low-melting alloys with DTA crucible materials, while the latter is very readily oxidized. It is highly desirable to calibrate the DTA apparatus using at least two different substances, for any given set of experimental measurements. The mean area of the melting endotherm and the freezing exotherm will probably give a more reliable value than the melting endotherm alone.

The calibration constant, K, can readily be obtained from the equation

$$(\text{Total peak area}) = K \times (\text{known enthalpy change})$$

Experimental enthalpy changes can then be calculated as follows.

$$\text{Experimental enthalpy change} = \frac{\text{Experimental peak area}}{K}$$

K has units of cm^2 J^{-1}.

TEMPERATURE CALIBRATION OF DTA APPARATUS

Because differential thermal analysers differ widely in their design, constructional materials and methods of temperature measurement, ICTA realized early on that it would be essential to agree on a set of transitions that could be used for temperature standardization.[16] Accordingly a committee was set up under the chairmanship of Dr H. G. McAdie. A number of substances undergoing phase transitions at temperatures up to 1000 °C were tested by 34 separate workers, operating in 13 different countries and using 28 different types of DTA apparatus. Bearing these facts in mind, the range of extrapolated onset temperatures reported[17] (see Table 4.2) is not surprising. A much higher degree of reproducibility can readily be obtained by any individual worker, using only one piece of apparatus, under standard conditions.

Table 4.2 also lists the accepted equilibrium transition temperatures for the processes which have been recommended by ICTA for the calibration of differential thermal analysers. The substances used are all reasonably stable and are readily available in a high state of purity. However, tested calibration materials

TABLE 4.2

Element or compound	Transition	Equilibrium transition temperature/°C	Extrapolated onset temperature (ICTA standards test)/°C
KNO_3	solid–solid.	127.7	128 ± 5
In	solid–liquid	165.6	154 ± 6
Sn	solid–liquid	231.9	230 ± 5
$KC10_4$	solid–solid	299.5	299 ± 6
Ag_2SO_4	solid–solid	424[a]	424 ± 7
S_1O_2	solid–solid	573	571 ± 5
K_2SO_4	solid–solid	583	582 ± 7
K_2CrO_4	solid–solid	665	665 ± 7
$BaCO_3$	solid–solid	810	808 ± 8
$SrCO_3$	solid–solid	925	928 ± 7

[a]Taken from Ref. 18; all other data taken from Ref. 17.

can be obtained from the U.S. National Bureau of Standards: NBS-ICTA Standard reference material 758 covers the temperature range 125–435 °C; material 759 covers the range 295–675 °C, and material 760 the range 570–940 °C.

For workers primarily concerned with organic materials, the present ICTA standards are of little help. This has led to various attempts to assess the suitability of certain organic compounds for use in temperature calibration.[18, 19] Some of these substances and their transition temperatures are listed in Table 4.3.

TABLE 4.3

Compound	Transition	Equilibrium transition temperature/°C	Extrapolated onset temperature/°C	Ref.
P-Nitrotoluene	solid–liquid	51.5	51.5	18
Hexachloroethane	solid–solid	71.4	—	19
Naphthalene	solid–liquid	80.3	80.4	18
Hexamethyl benzene	solid–solid	110.4	—	19
Benzoic acid	solid–liquid	122.4	122.1	18
Adipic acid	solid–liquid	151.4	151.0	18
Anisic acid	solid–liquid	183.0	183.1	18
2-Chloroanthraquinone	solid–liquid	209.1	209.4	18
Carbazole	solid–liquid	245.3	245.2	18
Anthraquinone	solid–liquid	284.6	283.9	18

ESTIMATION OF HEAT CAPACITY (SPECIFIC HEAT)

As mentioned previously, even when a substance heated in a DTA cell is not undergoing a physical or chemical change, the DTA curve is still usually offset from the theoretical base line. This is due primarily to the difference in heat capacities of the test and reference samples, which results in each lagging to a different extent behind the temperature of the furnace. Hence the magnitude of the offset temperature provides us with a means of estimating heat capacities. A detailed theoretical consideration of heat capacity measurements has been published by David.[20]

Experimental method

The DTA instrument is first operated under standard conditions with no sample in either test or reference crucibles. Any deviation of the DTA curve from the theoretical base line will then be due to deficiencies in the apparatus, which are assumed to be constant. A test sample is then placed in one crucible, leaving the other empty, and the experiment repeated. Because of the change in heat capacity resulting from the presence of the sample, a new offset temperature will be attained.

The difference between the blank and experimental offset temperatures can then be used to calculate the heat capacity of the test substance at a given temperature. This is illustrated in Fig. 4.4.

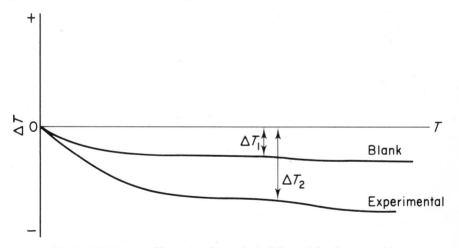

Fig. 4.4. DTA curves illustrating the method of determining heat capacities.

The value of the heat capacity at constant pressure, C_p, can then be evaluated using the following equation.

$$C_p = \frac{K'(\Delta T_2 - \Delta T_1)}{mr}$$

ΔT_2 and ΔT_1 are the offset temperatures in the presence and absence of a test sample, respectively; m is the sample mass and r the heating rate. The constant, K', must be obtained using a substance of known heat capacity. It should be noted that K' is only constant at a given temperature.

REPORTING DTA DATA

In order to avoid confusion and to ensure that sufficient information is given to enable work to be repeated, the ICTA Standardization Committee has drawn up a code of practice[21] for reporting thermal analysis data. These recommendations are now almost universally accepted; they should always be adopted, except, perhaps, where thermoanalytical measurements form only an insignificant part of a research report. So far as DTA is concerned, the ICTA code of practice calls for the following information to be given:

1. Identification of all substances; the sample, reference and diluent (if any).
2. A statement of the source of all these substances; their treatment and analysis.
3. Measurement of the average rate of temperature change.
4. Identification of the sample atmosphere by pressure, composition and purity and a statement as to whether the atmosphere is static, self-generated, or flowing through or over the sample.
5. A statement of the dimensions, geometry and materials of the sample holder.
6. Identification of the abscissa in terms of time or temperature. Time or temperature should be plotted to increase from left to right.
7. A statement of the methods used to identify intermediate and final products.
8. Faithful reproduction of all original records.
9. Wherever possible each thermal effect should be identified and supplementary supporting evidence stated.
10. Sample weight and extent of dilution of sample should be indicated.
11. Identification of the apparatus, including the geometry and materials of the thermocouples; the location of the differential and temperature measuring thermocouples.
12. The ordinate scale should be given as deflection per degree Centigrade at a specified temperature. Endothermic peaks should be indicated in a downward direction and exothermic peaks in an upward direction.

COMPLEMENTARY TECHNIQUES

It is seldom safe to rely on the evidence of DTA alone to elucidate a problem; techniques most commonly employed in conjunction with DTA are thermogravimetry (TG) and evolved gas detection (EGD). With the former, changes in weight of the sample, subjected to a uniform rate of rise of temperature, are recorded. This enables any DTA peak due to a chemical reaction involving a weight change to be identified. If a DTA peak occurs at a temperature where the sample weight remains constant, then it follows that either a solid state reaction, or a phase transition, must have occurred.

Evolved gas detection does not require a separate experiment to be carried out, as is the case with thermogravimetry. The DTA apparatus is used with a flowing inert atmosphere, and the effluent gas passed through one of the standard types of detector used for gas chromatography. Hence any DTA peak due to a process involving the evolution of volatile products, will have a corresponding peak on the EGD trace. Alternatively, if the DTA apparatus is being continuously evacuated, then EGD can readily be achieved by connecting a recording vacuum gauge into the pumping line. Evolution of volatile products will then lead to a transient increase in pressure, before they are removed by the pumping system.

If the evolved volatile products are not merely detected, but are analysed either qualitatively or quantitatively, then the procedure is known as evolved gas analysis, EGA. Samples may be collected at temperatures corresponding to DTA peaks, and subsequently analysed chemically, or by such techniques as infrared spectroscopy or gas–liquid chromatography. A more sophisticated method is to sample the effluent gas continuously, via a controlled leak into a mass spectrometer.

Many other complementary techniques have been used in conjunction with DTA, the choice depending very much on the type of substance being investigated. Among the physical properties that may be measured under conditions of a constant rate of change of temperature, are: length, volume, hardness, electrical conductivity, light reflectance or absorption and physical appearance by hot stage microscopy. A detailed discussion of those techniques falls outside the scope of this book, but they are well documented elsewhere (e.g. Daniels[22]).

SOURCES OF ERROR IN THE INTERPRETATION
OF DTA CURVES

So many factors are capable of giving rise to misleading DTA curves that it is impossible to discuss them all here. However, a brief mention of some of the more common problems likely to be encountered may prove helpful.

Sintering of a sample, which is otherwise physically and chemically unchanged, can lead to a marked reduction in thermal contact between the solid particles and the sample container. Subsequent heat changes undergone by the sample may be poorly transmitted to the container and hence to the detecting thermocouple.

Whenever a decomposition reaction results in the rapid evolution of gaseous products, there is always the danger of part of the sample being ejected from its container, due to the pressure of gas building up in the packed bed of particles. This is particularly prevalent when working with a continuously evacuated system. The ejected particles become deposited on various parts of the surrounding apparatus, where their subsequent thermal behaviour may cause a range of spurious effects on the DTA trace.

Inhomogeneity of a powder sample has already been referred to and is another major cause of lack of reproducibility in DTA curves. Packing density and the extent to which the sample container is filled also need to be controlled as far as possible.

Accurate positioning of the DTA sample containers, relative to the detecting thermocouples, is of vital importance and depends to a large extent on the design of the apparatus. Unless a closely similar thermal contact is obtained between the sample and reference material containers, and their corresponding thermocouples, serious base-line drift is likely to occur. Even slight displacement of either container during a DTA experiment will almost certainly invalidate the results.

Sample containers made of relatively inert metals, usually platinum, are now generally used. Nevertheless, the possibility that the container may catalyse, or even contribute to the overall decomposition reaction, must be considered. Where more reactive, e.g. aluminium, containers are used, there is in addition the danger that oxide film formation on the container may seriously affect heat transfer.

SUMMARY

The important part of a DTA peak is where the curve first deviates from the base line. The extrapolated onset temperature (see Fig. 4.4) is much less affected by changes in experimental conditions than the peak temperature, and generally gives closest agreement with literature values for the corresponding process.No unique physical significance can be attributed to the peak temperature; it normally represents neither the maximum rate of a process nor completion of the process.

The area under a DTA peak is proportional to the total enthalpy change giving rise to that peak. However, the proportionality constant is not independent of temperature, so the apparatus must be calibrated in the same temperature range as that to be used for the experimental measurements.

It is unwise to attempt to interpret DTA curves without taking into account data obtained on the same substance by other techniques.

REFERENCES

1.   M. D. Judd and M. I. Pope, *J. Inorg. Nucl. Chem.* **33**, 365 (1971).
2.   S. L. Boersma, *J. Amer. Ceram. Soc.* **38**, 281 (1955).

3.   H. T. Smyth, *J. Amer. Ceram. Soc.* **34**, 221 (1951).
4.   A. F. Frederickson, *Amer. Mineral.* **39**, 1023 (1954).
5.   H. Morita and H. M. Rice, *Anal. Chem.* **27**, 336 (1955).
6.   E. M. Barrall, II and L. B. Rogers, *Anal. Chem.* **34**, 1101 (1962).
7.   E. M. Barrall, II, *Thermochim. Acta* **5**, 377 (1973).
8.   R. C. MacKenzie, *Talanta* **19**, 1079 (1972).
9.   P. F. Kerr and J. L. Kulp, *Amer. Mineral.* **33**, 387 (1948).
10. H. J. Borchardt and F. Daniels, *J. Amer. Chem. Soc.* **79**, 41 (1957).
11. E. Sturm, *Thermochim. Acta* **4**, 461 (1972).
12. V. V. Vaidya and P. S. Nicholson, *J. Thermal Anal.* **5**, 637 (1973).
13. L. G. Berg, V. P. Egunov and A. D. Kiyaev, *J. Thermal Anal.* **7**, 11 (1975).
14. M. J. O'Neill, *Anal. Chem.* **36**, 1233 (1964).
15. E. M. Barrall, II and L. B. Rogers, *Anal. Chem.* **34**, 1106 (1962).
16. H. G. McAdie, *Thermochim. Acta* **1**, 325 (1970).
17. H. G. McAdie, *Proceedings of the Third International Conference for Thermal Analysis*,
     Vol. 1, Birkhauser Verlag, Basel, 1972, pp. 591–608.
18. B. Wunderlich and R. C. Bopp, *Thermochim. Acta* **6**, 335 (1974).
19. H. Kambe, K. Horie and T. Suzuki, *J. Thermal Anal.* **4**, 461 (1972).
20. D. J. David, *Anal. Chem.* **36**, 2162 (1964).
21. H. G. McAdie, *Anal. Chem.* **39**, 543 (1967).
22. T. Daniels, *Thermal Analysis*, Kogan, Page Ltd., London, 1973.
23. E. E. Marti, *Thermochim. Acta* **5**, 173 (1972).
24. N. A. Lange, *Handbook of Chemistry*, 10th Edn McGraw-Hill, New York, 1967.
25. M. J. Richardson and N. G. Savill, *Thermochim. Acta* **12**, 221 (1975).

CHAPTER 5
Phase Transitions

The determination of melting and boiling points has traditionally been used as a method of identification of organic compounds. Differential thermal analysis has the advantage that in the case of melting points, both the temperature and the latent heat of fusion can be determined in a single experiment. The use of DTA for such measurements has been established for more than 25 years[1] and has been developed to the point where, in the words of E. M. Barrall,[2] it no longer needs to be justified—only to be applied.

Conventional methods of melting point determination rely on observation of the sample, which is heated or cooled in a controlled manner. The recorded melting point therefore depends to a certain extent on the judgement of the operator, since no quantitative record of the measurement is obtained. As the impurity content of a specimen increases, so melting occurs over a progressively wider range of temperature, rather than at one distinct temperature, as is the case with a truly pure sample. Hence the difficulty in making visual melting point determinations increases correspondingly. Differential thermal analysis can be used to measure both the melting point and mole fraction of impurity present; this is discussed in detail on pp. 78–80.

Further advantages of DTA for the determination of phase transitions are, first, that very small samples weighing perhaps only a milligram can be used without loss in accuracy. Second, since the sample does not have to be observed during the experiment, measurements can be made at very high temperatures and/or pressures, using suitable apparatus.

SOLID–LIQUID TRANSITIONS

Here we shall be concerned only with samples that can reasonably be regarded as 'pure'; for impure samples, please refer to pp. 78–80. With melting point determinations, it can be advantageous to use a classical type of differential thermal analyser, where the differential thermocouple junctions are in direct contact with the test and reference samples. The problem of the temperature gradient between the sample, its container and the thermocouple junction is then eliminated. With such an apparatus, two approaches are possible. The test substance can be coated directly onto the junction of the thermocouple, by dipping it into a melt of the test material and cooling, prior to setting up the DTA apparatus. Alternatively, the test substance must be diluted with a large excess of the reference material, so that the heat capacity and thermal conductivity of the test and reference cells are as nearly matched as possible.

However, virtually all modern commercial differential thermal analysers work on the Boersma principle, where the test and reference samples are placed in matched containers which are in direct contact with the thermocouple junctions. Under these circumstances, it is highly desirable to encapsulate the sample under an inert atmosphere; this serves both to prevent vaporization and to keep oxidation to a minimum. Any increase or decrease in vapour pressure resulting from changes in temperature after encapsulation will normally be too small to have any detectable effect on the melting point of the substance.

It has been found that some substances, particularly metals, do not always form a uniform liquid layer in the sample container after melting. Instead they break up into one or more spherical droplets, which make poor thermal contact with the container, and hence gives rise to erratic and irreproducible DTA peaks. To avoid this, Barrall[2] has suggested that an inert metal disc, of a diameter only slightly smaller than the sample container, should be pressed down over the sample before encapsulation. (Mica can also be used.) A similar disc would, of course, have to be placed in the reference container. The sample should be as small as possible, consistent with the sensitivity of the apparatus, so as to minimize thermal gradients.

Experimental procedure

The conditions suggested for the determination of melting and freezing points are as follows:

1. Use as small a sample as practicable, which should if possible be encapsulated under an inert atmosphere.
2. A metal disc, of the same material as the sample container, should be pressed down over the sample, to ensure the formation of a uniform liquid film.
3. The DTA head should be surrounded by a static, inert atmosphere.
4. Providing the sample is small, an empty container can be placed in the reference cell.
5. The heating (or cooling) rate should certainly not exceed the usual 10 °C min^{-1} and a rather lower figure is desirable for accurate work.
6. If starting with a solid, first raise the temperature until the melting endotherm is complete, then cool at the same rate so as to obtain the freezing exotherm. The area under these two peaks should be identical if no decomposition has occurred; the converse procedure applies for a liquid sample.

Interpretation of results

A detailed discussion of the interpretation of DTA curves has already been given, so the reader should refer to pages 32 and 36. Considerable controversy still exists as to the best method of obtaining melting points from DTA peaks. It is, however, generally agreed that the two significant points on the curve are (a) the extrapolated onset temperature and (b) the peak temperature. Both of these values depend on the way in which the temperature plotted on the abscissa is measured. Therefore it is highly desirable to calibrate the apparatus over the temperature range within which the melting point is to be determined. Where possible, ICTA temperature standards should be used, as described on pp. 36–38.

Irrespective of whether extrapolated onset or peak temperatures are used, it is suggested that the mean values obtained from the heating and cooling peaks will give the most consistent results (see Fig. 5.1).

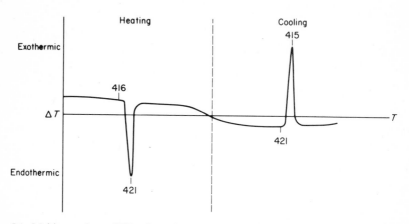

Fig. 5.1. Melting and resolidification of Analar zinc metal. Heating and cooling rates 5°C min^{-1} (Literature value for melting point of zinc 419.5°C).

The area under a melting endotherm is proportional to the latent heat of fusion of the test substance, so that once the apparatus has been calibrated in that temperature range (see pp. 36–38) the molar enthalpy of fusion can be calculated.

A single DTA curve will thus give both the temperature and enthalpy of fusion after an experiment lasting only a matter of minutes.

High-pressure measurements

Application of the DTA technique has made it possible to carry out melting point determinations under pressures of tens of thousands of atmospheres,[3, 4] thereby simulating the geological conditions during which igneous rock formation occurred. Since temperatures up to 2000 °C are involved, the apparatus has to be carefully constructed to exacting specifications, some details of which are given in the references cited. By measuring the variation of melting point as a function of pressure, information on the structure of solids can be obtained.

LIQUID–VAPOUR TRANSITIONS

It now seems to be generally agreed that data for liquid–vapour transitions obtained by differential thermal analysis or differential scanning calorimetry are more reproducible[2] than with classical methods of boiling point determination; this can be ascribed largely to the establishment of ideal conditions for equilibrium between the two phases. The use of DTA for the study of liquid–vapour transitions dates back to the mid-1950s,[5] since when numerous papers on the subject have appeared. Measurements have been made over a wide range of pressures, from less than 1 Torr to several atmospheres. Generally speaking, commercial differential thermal analysers can be used without modification, but some method for accurate control of the ambient pressure is essential. Unlike melting points, the temperature of a liquid–vapour equilibrium is markedly dependent on small changes in pressure, as described approximately by the Clausius–Clapeyron equation.

$$\log_{10}P = -\frac{L}{2.303RT} + C$$

P is the equilibrium vapour pressure at absolute temperature T, L is the molar latent heat of vaporization and R the gas constant (8.314 J mol^{-1}K^{-1}). C is an integration constant. Hence a graph of logarithm P plotted against $1/T$ should give a straight line, of slope $-L/2.303R$. This enables the molar latent heat of vaporization to be evaluated without attempting to determine the area under a boiling endotherm for a given mass of substance. Furthermore, since the true boiling point of a substance is the temperature at which its vapour pressure, under equilibrium conditions, reaches one standard atmosphere (760 mm Hg), the above graph can be extrapolated to give a more accurate boiling temperature.

However, three important assumptions have been made in deriving the Clausius–Clapeyron equation and these have a bearing on the accuracy of the data obtainable. First, it is assumed that the latent heat of vaporization is independent of temperature, which is not strictly true. Over a small temperature range, say 25 °C, errors from the source are not usually apparent; but if logarithm P is plotted against $1/T$ for a wide range of temperature, a smooth curve rather than a rectilinear graph is likely to be obtained. Secondly, the assumption is made that the volume of vapour formed is vastly greater than the volume of liquid producing it, so that the latter can reasonably be neglected. It is doubtful if any significant errors arise from this approximation. It is the third assumption, that the vapour formed on boiling of a liquid behaves as a perfect gas, which leads to real problems. In fact, deviations from ideal behaviour are normally greatest when a gas approaches its temperatures of liquefaction and there is no simple way in which corrections can be made.

Nevertheless, the limitations inherent in the use of the Clausius–Clapeyron equation apply not only to DTA, but to virtually all other methods normally used for the determination of boiling points and latent heats of vaporization. A recent evaluation of various techniques by Morie *et al.*[6] showed that an average relative error of 5.5% was obtained when using DTA data to calculate latent heats of vaporization.

A method claimed to be capable of greater accuracy has been put forward by Staub and Schnyder,[7] but, at the time of writing, insufficient information is available to assess the effectiveness of this technique. However, one undoubted advantage of their method is that it minimizes errors due to base-line drift, which result from changes in heat capacity of the sample consequent upon losses through evaporation, at temperatures well below the boiling point.

Experimental procedure

Two essential requirements for making accurate measurements are: (a) that the conditions necessary for the establishment of an equilibrium between the liquid and vapour phases must be provided and (b) that the ambient pressure must be controlled and recorded.

If the apparatus permits encapsulation of the sample, then single plate reflux conditions are readily obtainable by drilling a small hole, about 0.5–1.0 mm diameter, in the lid of the sample container. A metal disc insert, rather smaller in diameter than the container, can be placed over the sample. Where encapsulation is not possible, it will be necessary to coat the liquid sample over the surface of supporting, inert, solid particles. −200 BS mesh carborundum powder has proved very suitable for this purpose, with a liquid/solid ratio of not more than 2%. This has the effect of vastly increasing the surface area over which the liquid and vapour phases are in contact and minimizes superheating. An identical mass of carborundum powder must, of course, be used in the reference cell. Even where the sample can be encapsulated, supporting the liquid phase on carborundum powder, prior to encapsulation, is recommended.

Pressure control is a standard laboratory procedure which need not be discussed in detail here. It will normally be sufficient to use a continuously pumped system, with an inert gas admitted via a controlled leak, to obtain subambient pressures. Remember that it is necessary to measure the pressure in the vicinity of the DTA head itself, and not at some remote point in the pumping system.

The following procedure for making boiling point determinations is therefore suggested.

1. Use the smallest samples consistent with the sensitivity of the apparatus.
2. Coat the liquid on an inert powder, such as −200 BS mesh carborundum. Use an equal mass of carborundum as reference material.
3. Encapsulate the sample where practicable, with a hole 0.5–1.0 mm diameter in the top of the container.
4. Control the ambient pressure at the required value.
5. Do not use too fast a heating rate; a reasonable figure is about 5° min⁻¹ in the boiling region.

6. Record the extrapolated onset temperature of the boiling endotherm, as explained on p. 33. It is not possible to measure the latent heat of vaporization directly from the peak area, because the amount of sample which has evaporated at temperatures below the boiling point is unknown (see Fig. 5.2).

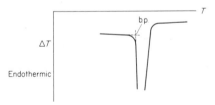

Fig. 5.2. A typical boiling endotherm; the change in baseline is due to evaporation of the sample.

7. Make further boiling point determinations over a range of pressures.
8. Plot a graph of the logarithm of vapour pressure against the reciprocal of the absolute temperature of boiling.

Interpretation of data

The graph plotted in point (8) above should be extrapolated so as to obtain the temperature at which the vapour pressure equals one standard atmosphere (760 mm Hg). This gives the boiling point of the liquid directly in K. To obtain the molar latent heat of vaporization, L, measure the slope of the graph, which is equal to $-L/2.303R$. Since the gas constant, R, has a value of 8.314 J mol^{-1}K^{-1},
$$L \text{ (J mol}^{-1}\text{)} = 19.10 \times \text{slope}$$
Do not attempt to obtain L from the area under a single endotherm, or even from average values, for the reason mentioned in point (6) above.

SOLID–SOLID TRANSITIONS

These fall essentially into two categories: (1) transitions where the change in crystal structure is accompanied by the evolution or absorption of heat and (2) transitions where the two phases have different heat capacities, but negligible heat is evolved or absorbed when one phase is converted to the other. Into the first category come many crystal transformations, such as that in quartz at 573 °C, which appear to be so reproducible that the magnitude of the heat change involved has been taken as a measure of the amount of that substance present. This is discussed on p. 80, where it is also pointed out that the authors have serious reservations about the validity of this method. It has been shown convincingly[8] that the shape of a DTA peak accompanying the quartz transition is drastically affected by grinding the sample. This can be ascribed to reduction in the crystallite size, or to lattice strain induced in the crystallites as a result of grinding; both factors may well contribute. Furthermore, Murat[9] has illustrated the wide variety of peak shapes and areas that can be obtained for the hexagonal–orthorhombic transition in anhydrite.

The investigation of type (1) transitions by DTA is a comparatively simple matter, since neither the pressure nor the composition of the ambient atmosphere have much effect on the resulting curve. Typical operating conditions would be a heating rate of $10°$ min^{-1}, a static inert atmosphere and a sample weight of 100 mg or less. The resulting peaks are often very sharp and narrow, so that peak temperatures, rather than extrapolated onset temperatures are frequently quoted. However, the former are still preferable. Providing the apparatus is

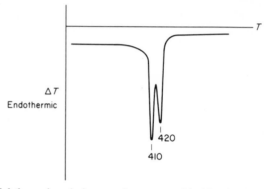

Fig. 5.3. Differential thermal analysis curve for cuprous chloride, showing a solid–solid phase transition endotherm, closely followed by melting.

suitably calibrated and operating conditions are carefully standardized, then the area under the DTA peak may be taken as a measure of the associated enthalpy change, or of the amount of substance present, subject to the reservations dis‑ cussed above. With very narrow peaks, height rather than area has sometimes ‑been used to obtain quantitative data.

With type (2) transitions no peak appears on the corresponding DTA curve. Instead the curve either shows a discontinuity at the transition temperature, or else undergoes a sudden change in slope. Many plastic–glass transitions fall into this category; these are considered in detail in the section on plastics and poly‑mers. The transition temperature is thus readily obtainable from the DTA curve, since the actual heat capacity values before and after the transition do not need to be known. Such measurements are frequently made at subambient tempera‑tures, so that controlled cooling is required. Both heating and cooling DTA curves are usually determined, to discover if the transition shows any hysteresis effects with change in temperature.

Interpretation of data

With type (1) transitions, interpretation of the DTA curve is a simple matter, since the phase transition peaks are generally sharp and occur at a highly repro-

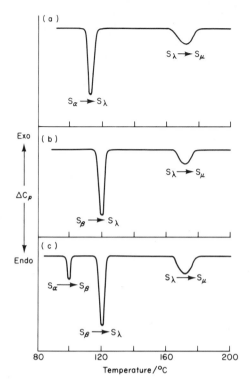

Fig. 5.4. Differential thermal analysis curves of: (a) single crystal of 6N sulphur (S_a); (b) pure S_β; and (c) microcrystalline S_a. (From Currell and Williams, *Thermochim. Acta* **9**, 255 (1974).)

ducible temperature. Providing the transition is reversible and does not show temperature hysteresis, a 'mirror image' peak should be obtained on cooling. Extrapolated onset temperatures, rather than peak temperatures, normally give the best measure of the transition temperature. However, with very narrow peaks, the latter are more readily determined.

Type (2) transitions may, or may not, be accompanied by a small enthalpy change, as illustrated in Fig. 5.5. In either case, the rectilinear parts of the DTA curve above and below the transition temperature should be extrapolated, as shown in the diagram. The actual transition temperature can then be obtained by finding the point at which a line midway between these extrapolated curves intersects with the DTA trace. This method will perhaps be more easily understood from an inspection of Fig. 5.5.

One factor which can have a pronounced effect on the recorded enthalpy changes accompanying solid–solid phase transitions, as measured by the DTA peak area, has only recently received serious consideration. Berg and his co-workers have shown[10] that the thermal conductivity of the gaseous atmosphere surrounding the solid particles undergoing a phase transition can lead to errors of as much as 25% in the observed enthalpy change. Correction of these errors

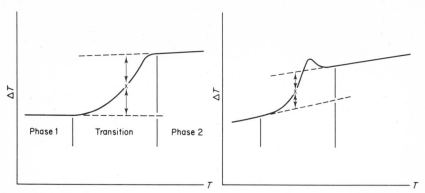

Fig. 5.5. Type II solid–solid phase transitions, (a) showing no evolution or absorption of heat and (b) accompanied by a small enthalpy change. The arrows indicate the transition temperature.

has already been discussed (see Chapter 4). However, serious errors are only introduced where the gaseous atmosphere has a relatively high thermal conductivity. In any case, this problem does not arise where both calibration and measurements are made in the presence of the same ambient atmosphere.

REFERENCES

1. M. Vold, *J. Anal. Chem.* **21**, 683 (1949).
2. E. M. Barrall II, *Thermochim. Acta* **5**, 377 (1973).
3. S. N. Vaidya and G. C. Kennedy, *J. Phys. Chem. Solids* **32**, 2031 (1971).
4. J. Akella, J. Ganguly, R. Grover and G. C. Kennedy, *J. Phys. Chem. Solids* **34**, 631 (1973).
5. H. Morita and H. M. Rice, *Anal. Chem.* **27**, 336 (1955).
6. G. P. Morie, T. A. Powers and C. A. Glover, *Thermochim. Acta* **3**, 259 (1972).
7. H. Staub and M. Schnyder, *Thermochim. Acta* **10**, 237 (1974).
8. G. S. M. Moore and H. E. Rose, *Nature* (*London*) **242**, 187 (1973).
9. M. Murat, *J. Thermal Anal.* **3**, 259 (1971).
10. L. G. Berg, V. P. Egunov and A. D. Kiyaev, *J. Thermal Anal.* **7**, 11 (1975).

CHAPTER 6

Phase Diagrams of Condensed Systems

Chapter 5 has already described the uses of DTA in the determination of phase transitions and has illustrated some of the experimental problems involved when using single pure compounds. This Chapter is concerned with the determination of phase transitions in multicomponent systems and the uses of DTA in the construction of phase diagrams.

Differential thermal analysis is a widely used technique for this type of study and offers many advantages over classical methods. The interpretation of the results obtained by DTA is no more complex than that needed by using any other method. In fact it does offer many distinct advantages. However, before we come on to discuss the relative merits of DTA and the problems associated with the technique, it is necessary to describe what shape and form the DTA results will take. In fact it is probably true to say that in the determination of phase diagrams, the shape of the DTA peak is of great importance and yields useful information.

This chapter, therefore, will be mainly devoted to the consideration of some hypothetical phase diagrams and will consider how the data obtained by DTA can be utilized. In order to make the reader aware of the basic physical chemistry behind phase diagrams it will be necessary to decribe this at some length. To those readers well versed in phase studies this may seem somewhat trivial but it is considered important for a complete understanding of what the DTA results really mean. Following on from this discussion, the relative advantages, disadvantages and problems of using DTA will be summarized. It is not proposed here to quote large numbers of examples since these only serve to demonstrate that DTA is widely used in phase studies. For specific examples Refs. 1–19 will serve as a good starting point.

Let us start by considering a simple two component system, where the components are completely miscible in the liquid state and the solid phases consist of pure components. If a liquid mixture of two components is cooled, solid will commence to separate at a definite temperature; this is the freezing point. At this temperature there are two phases, liquid and solid (if the vapour is ignored) and since there are two components the number of degrees of freedom can be calculated from the Phase Rule:

$$F = C - P + 2 = 2 - 2 + 2 = 2$$

There are, therefore, two degrees of freedom for such a system. However, it is common practice to look at such systems in the open atmosphere and there one degree of freedom, namely pressure, is fixed. In other words the temperature or the composition of the liquid phase will suffice to define the system fully. This can be explained in another way in that for every liquid mixture there will be a definite temperature at which it is in equilibrium with solid; this temperature is of course the freezing point. If the freezing points of a series of liquid mixtures, varying in composition from one pure component A to the other component B are measured and the results plotted against the corresponding liquid compositions, two curves represented by AC and BC will be obtained. This is illustrated in Fig. 6.1. When liquids rich in A are cooled, that is between

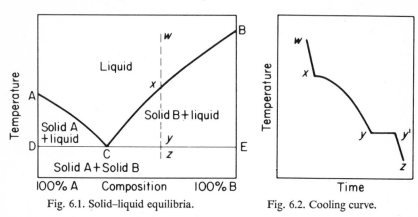

Fig. 6.1. Solid–liquid equilibria.   Fig. 6.2. Cooling curve.

A and C, solid A separates whereas liquids rich in B will deposit solid B on cooling. At the point C where the two curves meet there is the situation that solids A and B must be in equilibrium with the liquid. By application of the Phase Rule to this case where three phases co-exist (under constant pressure) it is found that the number of degrees of freedom available to the system is now zero. The system is now invariant. That is to say that there is only one temperature, at atmospheric pressure, where the liquid phase can be in equilibrium with both solids. This point C is the lowest temperature at which any liquid mixture of A and B will freeze and is called 'the eutectic point'.

Let us now examine the effects observed on the cooling of liquid mixtures of A and B. Consider, for example, a system of composition represented by the line wz. w represents the system entirely in the liquid phase and z when it is entirely solid. If the liquid w is cooled, no solid will separate until the point x on the freezing point curve BC is reached; at this temperature solid B will commence to deposit. The formation of solid will result in the liberation of the heat of fusion, so that at the point x the rate of cooling will be decreased. As the temperature continues to fall, the state of the system as a whole will be indicated by points x and y. When the eutectic point C is reached, the second solid A starts to deposit and since, as has already been described, there is only one temperature at which liquid and two solid phases can co-exist, the temperature must remain constant until all solid A has been deposited. The eutectic temperature is therefore marked by a complete stop in the rate of cooling and only when the whole system has solidified can the temperature fall further.

If a cooling curve was plotted for the above process it would resemble that given in Fig. 6.2. Point x is the point at which first crystallization occurs and point y is the eutectic temperature; obviously the closer the composition given by point x is to the eutectic composition, the closer will be points x and y on the cooling curve. Therefore by plotting a graph of points x and y against composition, the phase diagram may be constructed.

Let us now compare this cooling curve with the DTA trace obtained on a sample of the same composition cooled under the same conditions. From w to x the sample will cool at the same rate as the reference material. There may well be a pronounced slope of the base-line due to the difference between the heat capacity of the molten sample and a solid reference. At point x, solid B commences deposition and, as mentioned above, this results in the liberation of the heat of fusion. The temperature of the sample will therefore increase relative to that of the reference and an exotherm will appear on the DTA trace. This exotherm will, however, not be a single sharp peak but will be initially sharp and will tail off as complete crystallization occurs. Its overall size will be dependent on the amount of compound B present in the mixture. As the eutectic temperature is approached, the DTA trace will once again approach a stable base-line, as the temperature differences between sample and reference approach zero. At the eutectic temperature, solid A begins to crystallize and, in agreement with theory, the temperature of the sample must remain constant until all of component A has solidified. Since, of course, the reference material will still be cooling at the programmed rate, the eutectic temperature is marked by a sharp exotherm.

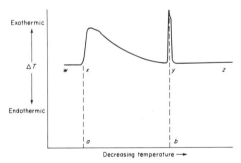

Fig. 6.3. Typical DTA curve obtained on a sample represented by w in Fig. 6.1.

After all of A has solidified, further cooling of the sample occurs and the DTA trace re-attains a stable base line. The DTA trace will, therefore have the appearance as shown in Fig. 6.3. If, of course, composition w were chosen to be that of the eutectic composition, the DTA trace would only show a single sharp exotherm. The phase diagram may therefore be built up in an analogous manner to that of using conventional cooling curves. In this case, the temperatures which should be plotted are the onset of the broad exotherm ((a) in Fig. 6.3) and peak maximum ((b) in Fig. 6.3.) The choice of this latter temperature, the temperature for a liquid–solid transition, is open to debate and has been discussed in the Chapter 5.

This, therefore, is the basic principle behind the use of DTA in the determination of phase diagrams and it should now be apparent as to the relationship between the cooling curves and a DTA trace. In fact, with DTA, in many cases, it is preferable to use the heating curve rather than the cooling curve in order to eliminate problems of supercooling. It goes without saying that in the example quoted above, if the sample were heated from x to w, endothermic peaks would be obtained on the DTA trace, which would be a mirror image of Fig. 6.3.

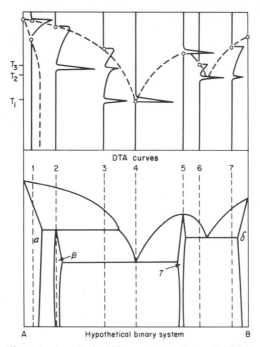

Fig. 6.4. Hypothetical binary system. (From Ref. 20.)

A more complex example of the use of DTA is given in Fig. 6.4. This represents a binary system exhibiting incongruently and congruently melting compounds and solid solution, eutectic and liquidus reactions. Typical DTA traces, under programmed heating conditions, at seven selected compositions, are given.

1. This shows only the α solution melting until the liquidus temperature is reached. The very sharp cut-off of the peak is always good evidence that the sample is all in the liquid phase.

2. The trace exhibits a sharp thermal arrest since the composition has been chosen to be that of the incongruently melting compound β. At the temperature of this endotherm (the peritectic temperature) the sample undergoes an isothermal decomposition. Above this temperature, melting occurs at an increasing rate until the liquidus is reached.
3. Here the composition is located on the eutectic side of the incongruent compound. The first reaction occurs when β and γ phases simultaneously melt at the eutectic temperature. Further heating causes the β phase to go into solution until a temperature is reached at which the β phase decomposes to give α plus liquid. Melting then continues until the liquidus is reached.
4. As has been described in the previous example, heating a mixture of the eutectic composition will lead to a single sharp endotherm.
5. This trace is similar to (4) and reflects the melting of the congruent compound γ.
6. and 7. These traces show the effects of the eutectic isotherm and continued melting until the liquidus temperature is reached.

Let us now attempt to construct the binary phase diagram from the information given by these seven DTA traces. The first step is to draw a curve through the temperatures of the highest temperature peaks on the traces; this will represent the liquidus curve. Having obtained the liquidus curve, then one should look for evidence of the presence of any isotherms, indicated by sharp peaks at the same temperature in different traces. Obviously in this example, there are three isotherms suggested at temperatures T_1, T_2 and T_3. Trace 3 has two isotherms suggested. The lower temperature peak is sharp and well defined and, since it is apparent in trace 4, strongly suggests the eutectic temperature. The higher temperature peak, because of its shape and size would strongly suggest a peritectic reaction. Therefore, with a few more thoughtfully located compositions it would be possible to define completely such a binary system as this.

What, therefore, are the relative advantages and disadvantages of using DTA for such an investigation, and what precautions should be taken with regard to sample preparation, etc. Consider the last point first.

SAMPLE PREPARATION

The preparation of the sample is crucial in determining the reliability of the method. In any two or multicomponent systems, the use of small samples for DTA may lead to sampling errors. Mixing up 1 g of mixture and taking 100 mg only will undoubtedly lead to large errors in composition. It is therefore preferable to mix the components directly into the DTA crucible. If this is not practicable then the components may be mixed on a macroscale, heated to above the melting point in an inert atmosphere, cooled to room temperature, ground to a fine powder, and a sample of this used for the DTA experiment. In many cases, however, where high temperatures are involved this may not prove practicable. From the above statements it should be apparent that the largest sample practicable should be used. Having prepared the sample, it should be placed in a suitable crucible. This crucible should be made of a material which is inert with respect to the sample under investigation. In the ideal case the sample should be encapsulated under an inert atmosphere to prevent oxidation or composition changes, where materials with a high vapour pressure are involved. If encapsulation is not possible, then conventional crucibles may be used, but it is advisable to maintain a static inert atmosphere around the sample. The final problem comes in the selection of the heating or cooling rate. This can only be determined by experiment and may well vary from system to system, depending on the sample under investigation, the equipment used, etc. In general, heating or cooling rates in the region of $5°$ min^{-1} or less are commonly used.

RELATIVE ADVANTAGES/DISADVANTAGES OF THE USE OF DTA

Advantages

DTA can be used up to high temperatures (2000 °C) and will detect any effect involving a heat change. At temperatures above 1000 °C it is virtually the only technique that can be used with any degree of certainty. Optical methods become difficult at these temperatures because of the light emitted by hot furnaces. Measurements below room temperature are possible utilizing low-temperature DTA equipment. Differential thermal analysis under constant pressures, other than one atmosphere, is also possible, and the variation of the temperature with pressure of certain transitions may yield much additional information. Special equipment has been built[21] which combines a DTA cell and a hot-stage microscope; this has been described in some detail in Chapter 2. Finally, DTA is in effect an automated system; once the experiment has been started the operator is free to do other work and this involves a considerable cost saving.

Disadvantages

There are very few disadvantages in using DTA for this sort of investigation. Conventional methods, especially those using cooling curves, are prone to many problems themselves. Perhaps the only drawback with DTA is that problems may be encountered when the sample starts to melt. As it melts it will contract away from the crucible walls and therefore thermal contact will be decreased. This will also lead to a marked difference in heat capacity between the sample and the reference material, with a consequent increase in base line drift. With modern equipment, however, this effect can be compensated for electronically.

CONCLUSION

This, therefore, gives a brief introduction into the use of DTA in the construction of phase diagrams. The discussion has been confined to two component systems for the sake of clarity, but obviously the technique is not limited to these. Also we have only described the more common types of system; nevertheless, it should be said that in theory every effect which involves a change in heat can be recorded by DTA, although considerable experience may first be necessary to interpret the results. The reader is strongly recommended to carry out the practical determination of a simple two component system (such as the $Na_2SO_4/NaCl$ system), before attempting anything more complex.

REFERENCES

1. J. Inoue, K. Osamura and K. Murakami, *Trans. Jpn Inst. Met.* **12**, 13 (1971).
2. A. R. West and F. P. Glasser, *J. Mater. Sci.* **6**, 1100 (1971).
3. D. T. Peterson, T. J. Poskie and J. A. Straatmann, *J. Less Common Met.* **23**, 177 (1971).
4. M. Polo, N. Gerard and M. Lallemant, *C. R. Acad. Sci. Ser. C* **272**, 642 (1971).
5. A. N. Campbell, A. N. and W. H. W. Wood, *Can. J. Chem.* **49**, 1315 (1971).
6. Y. Ozaki and S. Saito, *Jpn J. Appl. Phys.* **10**, 149 (1971).
7. J. P. Guha and D. Kolar, *J. Mater. Sci.* **6**, 1174 (1971).
8. G. Bruzzone, *J. Less Common Met.* **25**, 361 (1971).
9. A. S. Dworkin and M. A. Bredig, *J. Phys. Chem.* **75**, 2340 (1971).
10. A. K. Chatterjee and G. I. Zhmoidin, *J. Mater. Sci.* **7**, 93 (1972).
11. R. R. Dayal, *J. Less Common Met.* **26**, 381 (1972).
12. R. K. Datta and J. P. Meehan, *Z. Anorg. Allg. Chem.* **383**, 328 (1971).
13. F. Seel, K. Velleman and E. Heinrich, *Z. Anorg. Allg. Chem.* **382**, 61 (1971).
14. E. Rozen and R. Tegman, *Chemico Scripta* **2**, 221 (1972).
15. A. H. Schultz and B. Bieker, *Acta Chem. Scand.* **26**, 2623 (1972).
16. M. Samouël and A. Kozak, *Rev. Chim. Minér.* **9**, 815 (1972).
17. M. Laugt, *C. R. Acad. Sci. Sér. C* **275**, 1197 (1972).
18. J. P. Guha and D. Kolar, *J. Mater. Sci.* **7**, 1192 (1972).
19. R. H. Nafziger, R. H. Lincoln and N. Riazance, *J. Inorg. Nucl. Chem.* **35**, 421 (1973).
20. D. E. Etter, P. A. Tucker and L. J. Wittenberg, *Thermal Analysis, Proceedings of the 2nd International Conference for Thermal Analysis*, Worcester, U.S.A., Academic·Press, New York, 1969, p. 829.
21. R. P. Miller and G. Sommer, *J. Sci. Inst.* **43**, 243 (1966).

CHAPTER 7
Thermal Decomposition of Solids

The very first reported[1] use of differential thermal analysis, by Le Châtelier in 1887, was in connection with the decomposition of solids, in the form of clays. Since then, literally thousands of papers on this subject have appeared in various journals. However, most simple organic solids melt and frequently vaporize prior to decomposition, so that DTA has found little application in this field. Special cases like the fission of ligands from inorganic coordination compounds and the thermal decomposition of high polymers are dealt with in separate sections. Also considered elsewhere are such topics as explosives and propellants, and solid fuels.

This chapter is therefore concerned very largely with the decomposition of inorganic compounds, including hydrates, which do not melt below their decomposition temperature. Differential thermal analysis is now seldom used as the sole means of investigating a decomposition process, since it gives little or no information as to the chemical constitution of the decomposition products. Furthermore, it is not possible to be sure which DTA peaks correspond to a chemical reaction and which are the result of physical processes, such as melting, boiling or solid–solid phase transitions. An added complication is that some DTA peaks may arise through secondary reactions between evolved decomposition products; these can occur in the gas phase, or catalytically on the surface of the decomposing sample, or even on its container.

For these (and several other) reasons DTA is most commonly employed in conjunction with thermogravimetry (TG), so that the chemical constitution of any solid compounds formed during decomposition can be determined. While TG curves for virtually all common inorganic compounds have been determined and a comprehensive treatise on the subject written by Duval,[2] it is most unwise to compare one's own DTA curves with TG data from other sources. This is because the rates, and even the course, of decomposition reactions may be influenced profoundly by such factors as: the ambient atmosphere, heating rate, presence of trace impurities, crystal orientation, lattice strain, packing density and even the chemical nature of the sample container. The TG and DTA experiments must therefore be performed on samples from the same source and under as nearly identical conditions as it is possible to achieve. Only then can a meaningful comparison between the two sets of data be made. The applications of TG are fully described in a recent book[3] by Dollimore and Keattch; anyone proposing to use DTA to investigate the decomposition of solids would be well advised to consult this book, in conjunction with the present chapter.

Many of the factors which affect the shape of DTA curves and hence influence their interpretation have already been considered in Chapter 4. The particular relevance of these considerations to the decomposition of solids will be discussed below.

EFFECT OF VARYING THE HEATING RATE

It is now common practice to employ a heating rate of 10 °C min^{-1} when using DTA to investigate the decomposition of solids. This is usually quite satisfactory, but changes in the heating rate can lead to quite pronounced and sometimes unexpected changes in the DTA curve.

If the potentiometric recorder producing the DTA curve has a fixed chart speed, then increasing the heating rate will cause each peak to become progressively higher and narrower, since the area should remain constant. The extrapolated onset temperature can be determined much more reproducibly for peaks which occur over a narrower temperature range. However, other effects resulting from a change in heating rate are far more difficult to predict. One might expect that slowing down the heating rate would allow better resolution

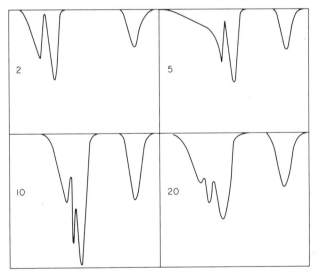

Fig. 7.1. Showing the effect of altering the heating rate from 2 to 20°C min^{-1} on the DTA curve for the decomposition of CuSO$_4$.5H$_2$O.

of overlapping, or closely spaced peaks. That the converse situation can sometimes occur is clearly illustrated in Fig. 7.1, which shows the effect of heating CuSO$_4$.5H$_2$O in static air, using 50 mg samples and heating rates of 2, 5, 10 and 20 °C min^{-1}. At the two highest heating rates, the first endotherm is resolved into three distinct peaks; these are due to the formation of $CuSO_4.3H_2O$, subsequent boiling of the water liberated and then decomposition of the trihydrate to form $CuSO_4.H_2O$. This latter compound remains stable up to a temperature of about 230 °C, when it decomposes to the anhydrous salt and gives rise to a separate, clearly defined, endothermic peak. If the heating rate is reduced to 5 °C min^{-1}, then the peaks due to formation of the trihydrate and boiling of the liberated water become merged, although a marked change in slope is apparent. At 2 °C min^{-1}, no trace of a separate endothermic effect due to boiling can be detected.

Hence, with this particular example, maximum information is provided by using a heating rate of 10 °C min^{-1}. Comparison of the curves for 10 and 20 °C min^{-1} suggests that the first two peaks would again merge if the heating rate was increased still further.

The highest acceptable heating rate will be determined primarily by the quality of the apparatus. However, the larger the sample used, the lower will be the maximum acceptable heating rate. This is due to the temperature gradient within the sample and the time taken for evolved decomposition products to diffuse to the surface.

In particular, the relatively small peaks due to solid–solid phase transitions can easily be lost, or merged with other thermal effects, as a result of using too high a rate of heating.

THE AMBIENT ATMOSPHERE

Except where it is desired to study a process such as dehydration of a salt in the presence of water vapour at a known, constant pressure,[4] a flowing inert atmosphere will normally be used. This is desirable for the removal of evolved decomposition products, which would otherwise retard the reaction. Hence better resolution will be obtained, due to sharper, more widely separated peaks occurring on the DTA trace. An added advantage of a flowing atmosphere is that the effluent gas can be analysed (EGA) either qualitatively, or quantitatively, to establish the nature and amount of the various reaction products (see Fig. 7.2).

There are several advantages[5] to be obtained from using a continuously evacuated system, where the nature of the information required allows this. Firstly, gaseous reaction products are removed much more rapidly and decomposition reactions often become detectable at much lower temperatures. Second, the evolved reaction products have little opportunity to undergo secondary reactions before they pass through a detector, or are 'trapped out' for subsequent analysis. Against this, it is more difficult to obtain uniform and reproducible heating rates in a continuously evacuated system; there is also the possibility of particles of the sample being ejected from the container by the pressure of escaping, gaseous, reaction products. If this happens, all manner of curious thermal effects may be observed, as particles decompose on various parts of the

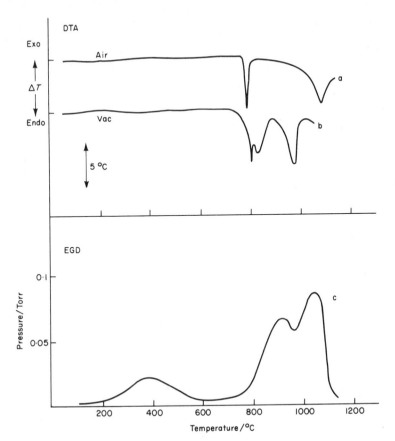

Fig. 7.2. Double (Sr–Ba) carbonate precipitated by ammonium carbonate. (a) DTA in air, (b) DTA and (c) EGD in a continuously evacuated system. Note the sharp endotherm due to the phase transition at 795 °C. (From Judd and Pope, 3rd ICTA, Vol. 2, pp. 777–792).

apparatus. However, the latter problem can be eliminated by working with single crystals, or very small samples.

Some work has been carried out on hydrates (and other materials) in sealed containers, so that the solid residue is confined in an atmosphere of its decomposition products, the pressure of which increases with temperature. Garn[6] found that some reactions became much more complex than expected under these conditions. For example, the apparently simple process:

$$BaCl_2 2H_2O \rightarrow BaCl_2 H_2O + H_2O(g)$$

actually gives rise to four distinct endothermic peaks. On the other hand, Wendlandt[7] in a study of various nickel salt hydrates, found that the DTA peaks were generally smaller and narrower, when the decompositions were carried out in sealed tubes (Fig. 7.3).

In calorimetric studies, a problem may arise through differences between the thermal conductivity of the atmosphere used during calibration and that of gaseous products surrounding the sample during decomposition. Calibration for calorimetric measurements is normally carried out (see p. 36) using the melting of pure substances, having melting points in the required temperature range

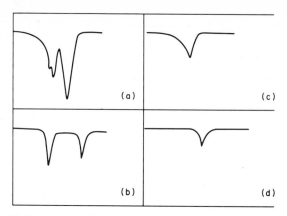

Fig. 7.3. Differential thermal analysis curves for (a) $CuSO_4.5H_2O$ in air (b) $CuSO_4.5H_2O$ in a sealed tube (c) $CuSO_4.3H_2O$ in air and (d) $CuSO_4.3H_2O$ in a sealed tube. (From Wendlandt, *Thermochim. Acta* **9**, 456 (1974).)

and an accurately known enthalpy of fusion. A static, inert atmosphere, possibly helium, may be used. During decomposition, however, the atmosphere temporarily surrounding the sample may well consist of such gases as carbon dioxide. Taking this example, Berg[8] has shown that an error of as much as 25% may be involved in heats of reaction determined by DTA. Errors from this source are only important where the thermal conductivities of the calibration atmosphere and that of the gaseous reaction products differ greatly. According to Berg, a correction can be made by means of the following equation

$$\frac{\text{Peak area in gas, } G_1}{\text{Peak area in gas, } G_2} = \frac{\lambda_s + \lambda_{G1}\left(\frac{\gamma_s}{\gamma} - 1\right)}{\lambda_s + \lambda_{G2}\left(\frac{\gamma_s}{\gamma} - 1\right)}$$

where λ_G and λ_s are the thermal conductivities of the gas, G, and the massive solid, S, respectively. γ_s is the crystal density of the solid, S, and γ is the bulk density of the sample.

SAMPLE SIZE

Generally speaking, samples used for thermal decomposition studies should be as small as practicable; some of the reasons for this are discussed in Chapter 3. An exception to this rule occurs where the sample material is not homogeneous, or contains discrete particles of an impurity. It is then necessary to use a larger sample, so that the results will be more representative of the bulk material. Small samples readily allow the escape of evolved decomposition products.

Where possible, there are many advantages to be gained from using single crystals; they should, however, be well annealed so as to avoid the mechanism of decomposition depending too much on lattice strain and the presence of dislocations. The danger of choosing a seriously unrepresentative sample must also be born in mind.

The question of sample dilution is another matter on which opinions differ. When in doubt, dilution is best avoided, even where the apparatus requires fairly large samples. The dangers of dilution arise mainly from the risk of reaction of the sample with the diluent, even though this may occur only to a minute extent. In addition, there is a risk of the diluent acting as a catalyst, either for the decomposition itself or for reaction between evolved decomposition products. Dilution will, however, be necessary in the study of explosives and highly exothermic reactions, so as to avoid risk to the apparatus—or worse!

SAMPLE CONTAINERS

The sample containers should be shallow, or only partially filled, thus allowing the ready escape of gaseous decomposition products. Fluidizing of the sample during decomposition is correspondingly minimized, so that ejection of particles will not occur.

It is obvious that the sample container must be constructed of material which does not react with the sample under the experimental conditions chosen. However, even such unreactive materials as quartz and platinum readily form compounds with certain fused salts. Quartz containers often lead to glass formation with inorganic oxides, while platinum forms relatively low melting point eutectics with various metals. (Do *not* heat lead salts in platinum buckets up to high temperatures.)

The catalytic action of platinum and certain other metals in promoting the decomposition of inorganic oxysalts is well known. An illustration of this effect is given in a recent paper by Morisaki and Komamiya,[9] who used both aluminium and platinum buckets for the thermal decomposition of ammonium perchlorate.

SOLID IMPURITIES—FLUXES

The presence of impurities which form low melting point eutectics with the host compound have been responsible for several misleading reports. Decomposition often effectively commences at the temperature when a liquid phase first appears; this frequently occurs well below the melting point of any of the single substances present. The action of fluxes is very well known but the possibility of such an occurrence affecting the thermal decomposition of solids is seldom considered. Hence, when the presence of an impurity is suspected, it is wise to consult the phase diagram for the system. Onset of decomposition at, or close to, the eutectic temperature will not then lead to false conclusions.

KINETICS AND MECHANISM

Although a separate section has been devoted to the use of DTA for studying the kinetics of reactions involving solids, certain points relevant to the decomposition of solids will be mentioned here.

The kinetics of solid state reactions are frequently described by a rate constant, k, and a reaction order, n. If decomposition proceeds by contraction of the reaction interface, the fraction of material decomposed, a, at time, t, is given by the following equation, where $n \neq 1.0$.

$$a = 1 - [1 - kt(1-n)]^{\frac{1}{1-n}}$$

Values of n equal to 0.50 and 0.66 correspond theoretically to the movement of a reaction boundary through disc, and spherically shaped, particles respectively. This rather simplistic view of matter has apparently led several workers to assume that only values of approximately 0.50 and 0.66 should be obtained. Studies on single crystals have clearly demonstrated that the reaction interface moves preferentially down dislocation lines, or regions of strain, giving rise to a highly irregular reaction zone, in no way resembling a circular disc or sphere. Accordingly, a whole range of values of n, not exceeding 1.0, are only to be expected in practice.

The second important point that needs to be made is that the temperature of the reaction zone is unknown. An exothermic reaction will raise the temperature of the reaction interface above that of its surroundings, and conversely for an endothermic process. If we make the very crude assumption that the rate of reaction doubles for each ten degrees the temperature increases, then the effect on the recorded kinetics of the temperature differential at the reaction interface will be apparent. This problem is serious enough in isothermal studies. Since DTA involves a uniform rate of rise in temperature, the difficulties in interpretation are correspondingly increased.

It must be concluded that DTA is not an ideal method for studying the kinetics of the decomposition of solids.

REFERENCES

1.   H. Le Châtelier, *C. R. Hebd. Séanc. Acad. Sci. (Paris)* **104**, 1443 and 1517 (1887); *Bull. Soc. Fr. Miner.* **10**, 204 (1887); *Z. Physik. Chem.* **1**, 396 (1887).

2.   C. Duval, *Inorganic Thermogravimetric Analysis*, 2nd Edn, Elsevier, Amsterdam, 1963.
3.   C. J. Keattch and D. Dollimore, *An Introduction to Thermogravimetry*, 2nd Edn, Heyden, London, 1975.
4.   V. Satava and B. Zbuzek, *Silikaty* **15**, 127 (1971).
5.   M. D. Judd and M. I. Pope, *Proceedings of the Third International Conference for Thermal Analysis*, Vol. 2, Birkhauser Verlag, Basel, 1972, pp. 777–791. See also *J. Appl. Chem. Biotechnol.* **21**, 149 (1971).
6.   P. D. Garn, *Anal. Chem.* **37**, 77 (1965).
7.   W. W. Wendlandt, *Thermochim. Acta* **9**, 101 (1974).
8.   L. G. Berg, V. P. Egunov and A. D. Kiyaev, *J. Thermal Anal.* **7**, 11 (1975).
9.   S. Morisaki and K. Komamiya, *Thermochim. Acta* **12**, 239 (1975).

CHAPTER 8
Chemical Reactions Involving Solids

The experimental techniques normally used to study reactions occurring in liquid, or in the gaseous phase, cannot in most cases be applied to reactions involving solids. This is because the system to be studied will not be homogeneous; reaction will occur only at the interface between two separate phases, rather than take place uniformly throughout a continuous, homogeneous medium.

Where a reaction involves changes in weight of the solid, as for example in oxidation or dry corrosion of metals, thermogravimetry (TG) provides an ideal means of following the course of the reaction, since the process can be studied isothermally. However TG cannot be applied to reactions where there is no overall weight change. Reactions of the type

$$nA \text{ (solid)} + mB \text{ (solid)} \rightarrow A_n B_m \text{ (solid)} - \Delta H$$

are of considerable industrial importance, representing the basic processes involved in the manufacture of glass, cements, ceramics, ferrites and certain catalysts, to quote but a few examples. In such cases, providing that the corresponding enthalpy change is reasonably large, DTA generally offers the most convenient method of studying the reaction. The only widely applicable alternative is to determine the X-ray diffraction pattern of the cooled reaction mixture after various periods of time. As the reaction proceeds, diffraction lines due to the reactants will gradually diminish in intensity, while a new series of lines due to reaction products will appear. Clearly such a technique will be time consuming, and because the reaction mixture normally has to be cooled before the X-ray diffraction pattern can be determined, phase transitions occurring during the cooling process will not be detected. Nevertheless, X-ray diffraction does provide a means of identification of the chemical structure and constitution of the reaction products. Accordingly, both X-ray diffraction and DTA are frequently employed together in the study of reactions involving solids; the former provides a means of product identification while the latter enables the course of the reaction to be determined.

It must not be thought that DTA is limited to reactions involving solids; the technique is equally applicable to reactions occurring in the liquid phase. In fact the classic paper by Borchardt and Daniels[1] is primarily concerned with the use of DTA to study the kinetics of decomposition of an aqueous solution of benzene diazonium chloride, and the reaction of dimethylaniline with ethyl iodide in the liquid phase.

SOLID–SOLID REACTIONS

The term solid–solid reactions is perhaps somewhat misleading, since it has been shown[2] that in many cases the presence of a transient liquid phase plays a vital part in the mechanism of reactions occurring between two or more solids. In the present context, solid–solid reactions are taken to include all cases where both reactants and products are in a solid state at room temperature.

A typical example of a reaction where thermogravimetry could not have been applied is the formation of bismuth titanate. The mechanism of this process has been studied by Solacolu, Cerchez and Segal,[3] who heated a mixture consisting of four moles bismuth oxide and one mole titanium dioxide in a DTA cell. The resulting DTA curve showed a broad exotherm occurring between 610 and 700 °C, together with two endotherms having peak temperatures at 730 °C and 865 °C, both the latter being reversible on cooling. The 730 °C endotherm can be ascribed to the well-documented phase transition occurring in pure bismuth oxide, while the 865 endotherm was shown to be due to melting of the reaction mixture. The 610–700 °C exotherm must therefore be the result of reaction between the two oxides: this was confirmed by pre-heat-treating the mixture at 725 °C, when subsequent DTA gave only the 865 °C endotherm. Hence no unreacted bismuth oxide remained in the reaction mixture and since titanium dioxide only melts at a much higher temperature, (c. 1850 °C) the 865 °C endotherm appears to be due to melting of the compound $4Bi_2O_3 . TiO_2$.

Experimental procedure

Before investigating the reaction between two solids, it is first necessary to determine how the individual materials behave on heating. This is because DTA of the reaction mixture will initially produce a curve showing a combination of all the exothermic and endothermic effects resulting from each of the materials present. Only peaks which do not occur with the single substances can be ascribed to a chemical reaction between them.

Where three (or more) individual substances are present in the reaction mixture, then it will also be necessary to study the thermal behaviour of each possible pair. Even though two substances may not react chemically, there remains the possibility that they may form a eutectic, melting at a much lower temperature than either of the pure components. If this occurs, then the overall phase diagram for this system will have to be determined, as described in Chapter 6.

Once the behaviour of the individual reactant, and pairs of reactants, has been established, an intimate and homogeneous mixture of the finely ground reactants has to be prepared, in the desired stoichiometric proportions. Generally it is unwise to grind the reactants together for any period of time, as this can lead to a premature tribochemical reaction.[4, 5]

The reaction mixture is then subjected to DTA in an inert atmosphere, with the heating process followed by controlled cooling, this cycle being repeated until no further evidence of reaction can be observed. During the first heating stage, peaks due to each individual reactant, together with peaks due to reaction between them, will appear. Any peaks on the heating curve due to phase transitions normally give rise to inverted (mirror image) peaks on cooling. Subsequent heating shows a progressive diminution in the size of the peaks due to the reactants, with a corresponding growth in peaks due to products. Eventually, equilibrium will be established, leading to a reproducible DTA curve being obtained.

Interpretation

The way in which curves due to reaction between solids are interpreted has already been indicated, but the procedure may become clearer if we consider a definite example. Wilburn et al.[6] have used DTA to investigate the complex reactions occurring during the manufacture of sheet glass. This is an extremely difficult subject to study and it is doubtful if the process could have been elucidated so successfully by any technique other than DTA. Sheet glass is manufactured by heating an intimate mixture of sand, sodium carbonate, dolomite and lime; accordingly each of these substances was examined separately by DTA. Sand merely showed the well known quartz transition endotherm at 573 °C. Sodium carbonate gave peaks at 100, 345, 470, 630 and 850 °C, the 100 ° peak corresponding to loss of water and decomposition of bicarbonate impurity. Peaks at 345° and 470° were associated with structural changes in the carbonate, which melted at 850°. Finally the peak at 630° was shown to be due to melting of a eutectic formed between the carbonate and sodium chloride present as an impurity. Dolomite gave peaks at 824° and 952°, corresponding to the decomposition of the magnesium and calcium carbonate components respectively; calcium carbonate in the form of limestone gave a decomposition peak at 975 °C.

Each possible pair of substances was then studied. Sand and sodium carbonate showed a combination of the individual peaks on heating; the glass forming reaction begins in the melt above 850 °C. On cooling an endotherm was observed at between 480° and 500 °C due to the product, soda glass. Sand and dolomite gave additional peaks at 1128° and 1194°, due to the formation and subsequent phase transition in Wollastonite, followed by melting at 1350 °C. Sand and limestone gave a reaction peak at 1442° due to the formation of dicalcium silicate; after cooling, reheating gave an endotherm at 662°, showing the presence of a glass. Sodium carbonate and dolomite, on reheating, gave peaks at 394°, and 437°, due to the formation of a double sodium calcium carbonate. Dolomite and limestone did not react.

From these data, it proved possible to interpret the DTA curve of the four component sheet glass mix. It appears that calcium enters the glass forming liquid in the form of the sodium calcium carbonate. The liquid forms due to heating sodium carbonate and silica above 900°, but as sodium metasilicate does not melt until 1088°, it follows that metasilicate must have reacted with more silica to form the disilicate; this gives a eutectic with silica which will melt as low as 780°. The higher the temperature of heat treatment, the greater the silica content of the glass becomes, causing the glass endotherm to rise from c. 500 to 585 °C.

The importance of melting in reactions apparently occurring between solids has been stressed by Berg.[2] He suggests that reaction commences by nucleation of the new phase on the parent solid, product formation occurring only slowly. A eutectic may then form between the reactant and product, leading to melting and rapid reaction at the solid–liquid interface. Clearly not all solid–solid reactions proceed in this way, but the necessity of knowing the phase diagram of the reactant–product mix has been demonstrated beyond doubt by the results of Berg and his co-workers.

A detailed consideration of the kinetics of reactions involving solids, with reference to DTA, appears elsewhere in this book. The first stages of the reaction tend to be controlled by the rate of nucleation of the new phase. These nuclei grow in size, until they combine to form a reaction interface moving through the solid particle. In practice, reaction tends to occur most rapidly along cracks and dislocation lines, where the structure of the reactant is distorted. This naturally leads to a most irregular reaction interface, and results in the kinetic parameters for the reaction being difficult to interpret theoretically.

SOLID–LIQUID REACTIONS

In principle, the use of DTA to study reactions between solids and liquids differs little from the study of solid–solid reactions. Indeed, solid–solid reactions frequently involve the formation of a liquid phase. Because of the volatile nature of liquids, it will often be necessary to work at elevated pressures to prevent evaporation. The desirability of providing the reaction vessel with a stirrer will also have to be considered. Another point worthy of mention is that relatively innocuous solids, e.g. alkali and alkaline earth halides, can form remarkably corrosive liquids, which readily attack DTA crucibles.[7]

Hence in practice, reactions involving liquids may often require a modified, or specially designed DTA cell. One field where DTA has been quite widely used is in the study of reactions between solids and liquid metals. The importance of such investigations lies in the use of liquid sodium and its alloys as a heat exchanging fluid in nuclear power generation. An apparatus specifically designed for such work has been described by Addison[8] et al., and is illustrated in Fig. 8.1. Steel reaction vessels (A) are used, 12 mm diameter and 40 mm in length, both

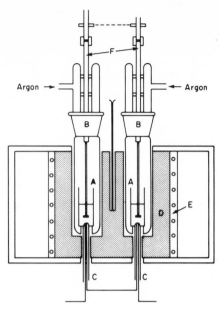

Fig. 8.1. The differential thermal analysis apparatus for the study of reactions between solids and liquid metals. (After Addison *et al.*, Ref. 8.)

contained inside silica tubes (B). A steel tube (C) is welded to the base of each reaction vessel and contains the thermocouple, insulated by silica sheathing. A drilled steel block (D) supports the silica tubes, the reaction vessels and their associated thermocouples. This block is heated externally and is capable of operating at up to 800 °C. Matched steel paddles serve as stirrers, their shafts sealed by O-rings, while a positive pressure of argon is maintained within the apparatus.

Interpretation of the DTA curves is essentially similar to that described for reactions between solids.

SOLID–GAS REACTIONS

Differential thermal analysis can be used to obtain information on commercially important solid–gas reactions using the same kind of apparatus as that designed for studying the catalytic activity of solids. Many processes in extraction metallurgy involve the oxidation, reduction, or chlorination of powdered mineral ores and are carried out on a very large scale.

In order to investigate the reactivity of a solid with respect to any given gas, the powdered sample is placed in a DTA cell of the type which permits the gas to flow through the packed bed of solid. Such a cell is illustrated elsewhere, in Fig. 11.1. The solid must then be outgassed to remove surface contamination;

this can be achieved either by heating the sample in a continuously evacuated system, or by heating in a stream of inert carrier gas. After cooling, the apparatus is flushed out with the reactant gas, such as oxygen or hydrogen, and the sample heated at a programmed rate of rise of temperature. The first exothermic deviation of the DTA curve from the base line will give the lowest temperature at which the reaction can be detected. The lower the reaction temperature, the greater the reactivity of the sample.

However, a conventional DTA cell is unlikely to be suitable for reactions involving chlorine, or other equally reactive substances, at elevated temperatures. Direct chlorination of mineral ores as a means of extraction is becoming increasingly important; recently, some Japanese workers[9] have succeeded in constructing an apparatus suitable for such investigations. This is illustrated in Fig. 8.2

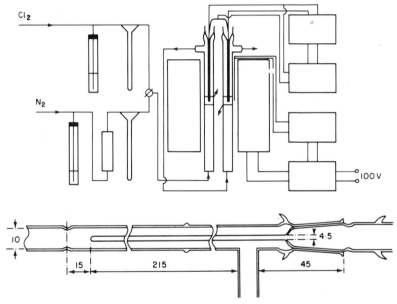

Fig. 8.2. Quartz DTA cell (unit of dimension: mm.) and associated apparatus. (After Ishi *et al.*, Ref. 9.)

and is capable of operation at up to 1000 °C. The apparatus consists essentially of two vertical silica tubes, 10 mm internal diameter, mounted side by side in a programmable furnace. Powdered sample and reference materials are held in position by wads of silica wool; closed silica tubes, 4.5 mm outside diameter, protect the thermocouple junctions located near the centre of each bed of powder. Flow rates for chlorine gas through the sample (*c.* 5 g) of up to 300 ml min^{-1} could be obtained.

Using this apparatus, the Japanese workers were able to show that the magnesium content of various ores could only be chlorinated by gaseous chlorine

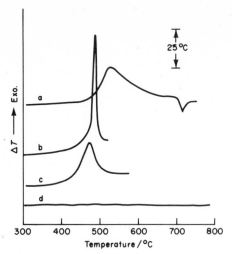

Fig. 8.3. Differential thermal analysis curves for the chlorination of MgO–C systems. C/MgO (molar ratio): curve (a) = 1; (b) = 2; (c) = 3; (d) = 0. Flowrate of Cl_2,35 ml min^{-1}; heating rate, 5°C min^{-1}. (After Ishi *et al.*, Ref. 9.)

when the ore was mixed with carbon. The effect of varying the carbon–magnesium oxide ratio was also studied; an example of the kind of DTA curves they obtained is shown in Fig. 8.3.

REFERENCES

1. H. J. Borchardt and F. Daniels, *J. Am. Chem. Soc.* **79**, 41 (1957).
2. L. G. Berg, N. P. Burmistrova and N. I. Lisov, *J. Thermal Anal.* **7**, 111 (1975).
3. S. Solacola, M. Cerchez and E. Segal, *Rev. Roum. Chim.* **18**, 203 (1973).
4. R. Albrecht, H. Hausler and R. Mobius, *Z. Anorg. Allg. Chem.* **382**, 177 (1971); *Z. Anorg. Allg. Chem.* **384**, 211 (1971).
5. H. Hausler, R. Mobius and P. E. Nau, *Z. Anorg. Allg. Chem.* **386**, 270 (1971).
6. F. W. Wilburn, S. A. Metcalfe and R. S. Warburton, *Glass Technol.* **6**, 107 (1965).
7. N. P. Burmistrova and R. G. Fitzeva, *J. Thermal Anal.* **4**, 161 (1972).
8. C. C. Addison, M. G. Baker and J. Bentham, *J. Chem. Soc. Dalton* 1035 (1972).
9. T. Ishi, R. Furvichi and Y. Kobayashi, *Thermochim. Acta* **9**, 39 (1974).

CHAPTER 9

Materials Specification, Purity and Identification

The use of DTA in both qualitative and quantitative analysis is widely established and many examples are referred to elsewhere in this book, in relation to particular groups of materials. In the present section, the various types of application will be discussed under a number of separate headings, together with their limitations and advantages. Generally speaking, DTA is more rapid than alternative methods of obtaining comparable information and uses only very small samples, or the order of milligrams, except where the material is inhomogeneous. Larger samples then become necessary, to avoid the risk of obtaining unrepresentative results.

A SIMPLE 'PASS/FAIL' ACCEPTANCE FOR TEST MATERIALS

A problem very widely encountered in industry is the necessity of establishing whether a recently supplied batch of material is identical with allegedly similar material, which had hitherto been used successfully. A rapid answer is usually necessary, while detailed analytical information is not required, at least for the time being. Differential thermal analysis is often ideally suited to solving this kind of problem.

The procedure is to place a weighed sample of known 'good' material, from a previous batch, in the reference cell of a differential thermal analyser. A similar mass of the suspect material is placed in the test cell and the apparatus heated (or cooled) at a uniform programmed rate, using a static, inert atmosphere. If the two samples are indeed identical, a rectilinear DTA trace should be obtained, with little or no base-line drift. The appearance of any peaks on the curve confirms at once that the two materials differ, either qualitatively or quantitatively. Serious base line drift also gives strong grounds for suspicion, since it suggests that the thermal conductivity, or heat capacity of the samples has not been matched, despite using similar weights.

Where one or more peaks appear on the DTA curve, then the temperatures at which these occur will very often give a good indication of the compounds responsible. Even small amounts of an impurity will often show up, due to lowering of the melting point of the host compound.

Exactly the same technique can be used to confirm the identity of a substance; for example, where the label on the container has been lost or defaced. The range of materials to which it can be applied is much wider than with the well-known 'mixed melting point' method, since the substance need neither melt, nor decompose chemically.

Liquids, as well as solids, can be investigated using this procedure, but in the former case controlled cooling (rather than heating) of the DTA apparatus will be necessary. Once the chemical composition of an impurity has been identified, then the amount present can often be estimated from the area under one of the corresponding DTA peaks. In order to do this at all accurately, it will be necessary to calibrate the apparatus by using standard mixtures, containing a range of known concentrations of the impurity concerned (see below).

THE 'MIXED MELTING POINT' METHOD OF COMPOUND IDENTIFICATION

This time-honoured procedure is worthy of a brief mention here, because it can readily be carried out using a DTA apparatus. The method involves guessing the identity of the unknown compound, after determination of its melting point. A small proportion of a pure sample (of known identity) of what is thought to be the same compound is then mixed with the original material; the mixture is then melted to ensure thorough mixing and allowed to re-solidify on cooling. On re-determining the melting point, no depression will be observed if the two compounds are in fact identical.

This method assumes that addition of an impurity which is completely miscible with the host compound in the liquid state, will always cause a depression of freezing point due to eutectic formation.

The identity of the compound, if guessed correctly, can therefore be determined from two DTA curves, covering the temperature range within which melting occurs. Any impurities present will lead to a broadening of the melting endotherm, as will be discussed below (see p. 78), so that the DTA curves will also give a qualitative indication of sample purity. Although the mixed melting point method is usually associated with low melting point organic compounds, use of DTA enables the same procedure to be applied over a much wider temperature range. Hence liquids can be studied by cooling them below their freezing points, while many inorganic compounds melt at temperatures readily obtainable by commercial differential thermal analysers.

One serious disadvantage of this method is that if the original, unknown material is already a mixture, rather than a compound, a great deal of time can be wasted. However, the shape of the DTA curve should provide clear evidence of the presence of a mixture (see Chapter 4).

QUANTITATIVE DETERMINATION OF PURITY FROM DTA PEAK AREAS

Very often, the presence of an impurity shows up as a separate peak on a DTA curve, in a temperature region where the host compound is thermally stable,

When this impurity peak (or peaks) is the result of a solid state thermal decomposition, or a solid–solid phase transition, then the following method can be applied. However, in the case of solid–solid phase transition peaks, first read pp. 80–82.

If the chemical nature of the impurity is known, then all that is necessary is to calibrate the DTA apparatus for that particular mixture. With inorganic compounds, standard mixtures can normally be prepared by intimate grinding of the two materials. Known amounts of each material are weighed out accurately, so as to give mixtures covering the required concentration range. A DTA curve for each mixture is then obtained, so that a calibration graph of peak area against impurity content can be drawn up. A knowledge of the enthalpy change for the corresponding reaction is not required.

Preparing homogeneous mixtures of organic compounds is not always possible by grinding. In such cases, an alternative method is to dissolve the two compounds separately in a volatile solvent. Mixtures of known concentration can then be prepared by quantitative addition of the two standard solutions, followed by evaporation of the solvent. This method does however carry the risk that the composition of the resultant mixture can be affected by fractional distillation of the two components during the evaporation process.

An example of the application of this technique is provided by a study of 'Spec-Pure' strontium carbonate, once used by the authors.[1] Differential thermal analysis of this material under vacuum gave rise to two endothermic peaks,

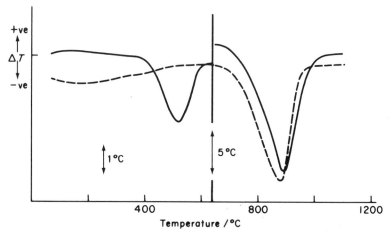

Fig. 9.1. Differential thermal analysis curves for Spec Pure strontium carbonate before (—) and after (– – –) treatment in carbon dioxide to remove traces of strontium hydroxide. (From Judd and Pope, Ref. 1.)

having extrapolated onset temperatures at 480 °C and 727 °C (see Fig. 9.1). Previous work,[2] including spectroscopic measurements, left little doubt that the lower temperature endotherm was due to the decomposition of strontium hydroxide. Accordingly any hydroxide present in the 'Spec-Pure' carbonate was decomposed by heating at 750 °C in a flowing carbon dioxide atmosphere, followed by cooling in the same atmosphere. Subsequent DTA (broken line, Fig. 9.1) showed that the hydroxide peak had been eliminated completely. The proportion of strontium hydroxide present in the original material could then be determined, after calibration with mixtures of the carbon dioxide treated strontium carbonate and pure strontium hydroxide.

QUANTITATIVE DETERMINATION OF SMALL AMOUNTS OF AN IMPURITY BY DEPRESSION OF MELTING POINT

An impurity which forms a eutectic with the host compound causes a lowering of the melting point, which is proportional to the mole fraction of the impurity present. This effect forms the basis of the classical method of molecular weight determination, where successive small amounts of the test solid are added to an inert solvent. The freezing point of the dilute solution is determined between each addition, and a graph plotted of the depression of freezing point against the mass of solute present. From the slope of the resulting line, the depression per gram of solute and hence its molecular weight can be calculated.

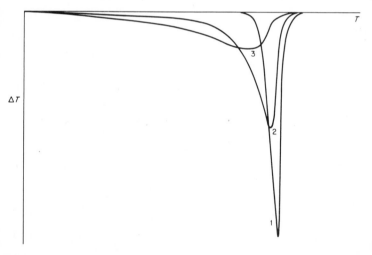

Fig. 9.2. Showing the effect of increasing impurity content on the shape of the DTA melting endotherm. Curve (1), most pure; (3), least pure.

However, addition of an impurity not only depresses the freezing (melting) point of a compound but also extends the temperature range over which freezing (melting) occurs. This is clearly demonstrated by comparing the DTA curves obtained during the melting of a compound, successive samples of which contain an increasing amount of an added impurity (see Fig. 9.2).

Calculation of the mole fraction of an impurity by depression of melting point relies on the thermodynamic equation,

$$\Delta T = \frac{RT_0^2}{\Delta H_f} x_2$$

where ΔT is the depression of melting point due to the presence of a mole fraction x_2 of the impurity. T_0 is the melting point of the pure compound (in degrees Kelvin), R the gas constant (8.314 J mol^{-1} deg^{-1}) and ΔH_f the molar enthalpy of fusion of the pure compound. In the derivation of the above equation, the following assumptions have been made: (1) the two components present form a eutectic phase diagram; (2) melting takes place at constant pressure; (3) the impurity forms an ideal solution; (4) the solution is dilute and (5) the enthalpy of fusion is independent of temperature. The extent to which all these assumptions are justified is often unclear in much of the published literature. Perhaps even more important is the difficulty in determining the depression of melting point, when melting occurs over an appreciable temperature range. This problem has been alleviated, if not solved, by using the modified equation shown below:

$$T_0 - T = \frac{RT_0^2}{\Delta H_f} \frac{x_2}{F}$$

Here T is the temperature of the solid–liquid equilibrium, when the fraction of the solid which has already melted is F. The value of F at any given temperature can be calculated from the ratio of the area under the DTA peak below temperature $T(a)$, over the total area of the melting endotherm (A) shown in Fig. 9.3. Hence a graph of T plotted against $1/F$ should give a straight line of slope $-RT_0^2x_2/\Delta H_f$ and intercept with the ordinate of T_0. Since the enthalpy of fusion of the pure compound needs to be known, this is best determined in a separate experiment using a high purity sample. The area under the melting

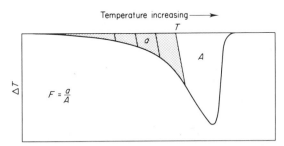

Fig. 9.3. Differential thermal analysis melting endotherm for an impure substance. The fraction melted (F) at temperature T is given by the shaded area (a) the total peak area (A).

endotherm can be used to calculate ΔH_f while the extrapolated onset temperature will give the value of T_0. Also the slope of the leading edge of the endothermic peak, s, gives a measure of the thermal lag in the system. Therefore in calculating the temperature, T, at which a given fraction (a/A) of the solid has melted, a series of lines are drawn parallel with a slope of s; the points of intersection of these lines with the abscissa give the corresponding values of T (Fig. 9.3).

The procedure described above is still only approximate and several papers have appeared describing more sophisticated (and highly complex) methods of data evaluation, often requiring the use of a computer.[3] A truly comprehensive review of this field has been published by Marti,[4] with particular reference to DSC, although most of the discussion is equally relevant to DTA. Marti concludes that the depression of melting point technique is capable of high accuracy for samples greater than 99 mol % purity, but that the accuracy falls significantly with increasing impurity content. Even more recently, this subject has again been reviewed by Palermo and Chiu,[5] with particular reference to the application of computational techniques.

Optimum experimental conditions for melting point determination are discussed elsewhere; they include a small sample mass, a slow rate of heating and a static, inert atmosphere.

PURITY DETERMINATIONS INVOLVING SOLID–SOLID PHASE TRANSITION PEAKS

Several papers have appeared which describe methods of determining the concentration of a substance from the height, or area under a phase transition peak, occurring at a characteristic temperature. Quartz, with its well authenticated solid–solid phase transition occurring at 573 °C, has been a subject of several investigations. Because of the chemical inertness of quartz, DTA provides an attractive alternative to the interpretation of X-ray powder diffraction patterns, often complicated by the presence of silicates. El Ghany et al.[6] used the height of the 573 °C endotherm to calculate the free silica content of ceramic raw materials. Calibration was achieved with standard mixtures of quartz in calcined alumina. A more sophisticated approach by Rowse and Jepson[7] enabled silica contents of up to 4% in clays to be determined, again using peak height as a measure of the amount of silica present. The results were verified by X-ray diffraction studies; a method of data evaluation was also described.

However, a subsequent paper by Moore and Rose[8] casts serious doubt on the general validity of this means of silica determination. They showed that the 573 °C endotherm was progressively reduced in height as a result of grinding and that it disappeared altogether after 400 h. Nevertheless, X-ray diffraction proved that the silica remained 75% crystalline and that the phase transition still occurred. After a comprehensive study, they concluded that the disappearance of the 573 °C endotherm was due to peak spreading; the normally sharp peak could in fact extend over as much as 50 °C. By using a greatly reduced heating rate, the phase transition endotherm could be made to reappear. Ruzek[9] has investigated the energy imparted to quartz by grinding and has shown that the exothermic release of this energy during DTA can reduce the size of the phase transition peak. Even the temperature at which this characteristic phase transition peak occurs does appear to be influenced by the geochemical history of the sample, according to the work of Takashima.[10]

It is therefore clearly apparent that the estimation of silica by this means must be treated with caution. Where the samples to be studied have an essentially similar history and composition, then the method may well be acceptable. Even then it is obviously desirable to check the validity of the results obtained by X-ray analysis.

These problems are certainly not confined to the α–β quartz transition, and probably occur to a greater or lesser extent with most solid–solid phase changes. Murat[11] has determined DTA curves for the hexagonal–orthorhombic transition in anhydrite, with fifty different samples: the wide variety of peak shapes and areas obtained illustrate only too well the dangers inherent in using solid–solid phase transition peaks as a means of quantitative analysis.

However, where the samples have all been prepared in a comparable manner, DTA sometimes provides a very convenient means of determining the extent to which one compound has been substituted in the lattice of another. For example, the authors have shown[12] that the amount of barium carbonate present in co-precipitated barium strontium double carbonate, used in the preparation of electron emissive coatings, can be determined accurately from the temperature of the orthorhombic →hexagonal phase transition. The variation of transition temperature with phase composition for this system is illustrated in Fig. 9.4; the temperatures recorded are those of the onset of the DTA endothermic peaks.

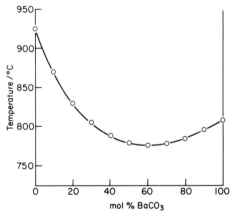

Fig. 9.4. The variation in peak onset temperature with composition of the phase transition in barium strontium double carbonate. (From Judd and Pope, Ref. 12.)

Yet another example of this type of application is provided by the work of Currell and Williams,[13] who determined the S_α content in samples of elemental sulphur, from the area under the endothermic peak occurring at 100 °C, due to the $S_\alpha \rightarrow S_\beta$ transition.

To sum up, it should be borne in mind that enthalpy changes accompanying a solid–solid phase transition appear to be less clearly defined than enthalpies of fusion, or vaporization. Lattice strain in a crystal, resulting from its previous history, can also influence the enthalpy change. Therefore the height of a DTA peak due to a solid–solid phase transition may often not provide an ideal means of analysis.

POLYETHYLENE STRUCTURE FROM MELTING ENDOTHERMS

Many organic polymers consist of a mixture of linear and cross-linked molecules, the two species having distinctly different melting points. This is well illustrated by polyethylene,[14] where the 'low density' polymer comprising almost entirely of cross-linked molecules, has a melting endotherm peak temperature of about 115 °C. The 'high density' material, in which the molecules are nearly all linear, does not melt until approximately 135 °C.

Where both linear and cross-linked molecules are present in the same material, the two peaks[15] appear on a DTA curve. If it is assumed that the enthalpy of fusion is the same for the two molecular structures, then the relative proportions of the species present can be determined directly from the ratio of the areas under the corresponding endothermic peaks. However, the procedure is not entirely straightforward, in that suitable pretreatment, such as annealing, is necessary to obtain clear peak separation and reproducible DTA curves. The reasons for this have been suggested in a more recent paper by Wunderlich,[16] who considers it quite possible for some polyethylene molecules to melt at a temperature as low as −10 °C. He claims that most polyethylene molecules melt within the range 80–135 °C, while some high molecular weight, extended chain molecules melt at 141.4 °C. It was further proposed that superheating may even raise the maximum melting point to 160 °C.

A more detailed discussion of the DTA of organic high polymers is given elsewhere in this book.

MATERIAL IDENTIFICATION BY DTA CURVE SHAPE

With complex materials, such as clays, silicates, coals and other organic polymers, both natural and synthetic, it is often impossible to provide a complete interpretation of all the peaks occurring in a DTA curve. Nevertheless, the curve obtained is often sufficiently characteristic of the substance concerned to serve as a means of identification, in a comparable manner to the infrared absorption spectrum.

The earliest application of DTA, reported by Le Châtelier in 1887,[17] was in fact used to classify clays into five types, according to their heating curves. For more than fifty years subsequently, DTA was largely confined to the identification of clays and silicates and to the estimation of one compound in the presence of another. Even quite small variations in structure, due to substitution of certain lattice ions by an impurity species, will often show up clearly in a DTA curve. The advantage of DTA over X-ray diffraction in this respect was recognised some time ago,[18] while the quantitative determination of clays and minerals by use of DTA had become a practical possibility by 1941.[19] Differential thermal analysis still remains one of the most important techniques for the identification and study of clays and related minerals; a comprehensive treatment of this topic can be found in the book[20] by Mackenzie.

Application of DTA to the identification of coals is of more recent origin; sufficient work had appeared by 1955 to justify the British Coal Utilization Research Association publishing a review[21] of this subject.

Plastics and polymers now provide probably the biggest field of application for DTA, although compound identification plays only a minor role. Several books concerned with this subject have been published in recent years.[22-25]

Various attempts have been made to establish a standard set of reference curves for the DTA of common substances. Mackenzie has compiled a punched card index for the curves of various minerals and inorganic compounds. A set of standard DTA curves for organic compounds is obtainable from the Sadtler Research Laboratories, Inc., 3316 Spring Garden Street, Philadelphia, Pa. 19104, U.S.A. More recently, a series of volumes under the general title, *Atlas of Thermoanalytical Curves*, has been published by Heyden and Son, Limited, under the editorship of G. Liptay.

IDENTIFICATION OF ORGANIC COMPOUNDS BY DERIVATIVE FORMATION

The identification of organic compounds by the formation of derivatives having known melting points is very widely used. Use of DTA for this purpose was suggested by Chiu[26] in 1962, who pointed out that it offered the advantages of being less time consuming, using very small samples and giving more information than a simple melting point determination.

The procedure proposed by Chiu involves first determining the DTA curve of the test substance, under static nitrogen, using a sample mass of 1–5 mg. Peaks on the resulting curve will indicate the melting point, boiling point, or any other phase transitions which occur over the temperature range covered. A heating rate of 10 °C min^{-1} and a maximum temperature of 500 °C will be suitable for studying practically all organic compounds. Evidence from the DTA curve will give a strong indication of the identity of the compound.

Next, an appropriate reagent is mixed with the organic test compound, with the more volatile substance present in excess. On re-heating in the DTA apparatus under the same conditions, peaks will appear due to reaction between the two materials, followed by the melting endotherm of the derivative. If finally the mixture is again heated, only peaks associated with the derivative should appear; the more volatile substance, formerly present in excess, will either have reacted or evaporated.

An example given by Chiu is the reaction between acetone and *p*-nitrophenyl-hydrazine, to form the *p*-nitrophenylhydrazone derivative; this is illustrated in Fig. 9.5. Acetone itself gives only the boiling endotherm, with an onset tempera-ture of 53 °C (curve a). *P*-Nitrophenylhydrazine (curve b) gave a melting endo-therm only, with an onset temperature of 147 °C. A mixture of the two substances shows various reaction peaks, together with volatilization of excess acetone (curve c) between 54 and 90 °C, followed by melting endotherm of the derivative.

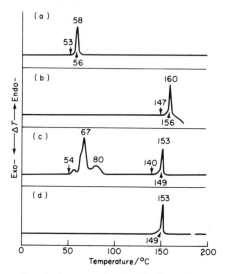

Fig. 9.5. Differential thermal analysis curves showing the formation of the *p*-nitrophenylhydra-zone derivative of acetone. (From Chiu, Ref. 26.)

Then in curve d, only the melting endotherm due to the characteristic *p*-nitro-phenylhydrazone derivative appears.

This technique should be generally applicable wherever compound identifica-tion by derivative formation is carried out.

REFERENCES

1.   M. D. Judd and M. I. Pope, *J. Appl. Chem.* **20**, 384 (1970).
2.   M. D. Judd and M. I. Pope, *J. Appl. Chem.* **19**, 191 (1969).
3.   S. A. Moros and D. Stewart, *Thermochim. Acta* **14**, 13 (1976).
4.   E. E. Marti, *Thermochim. Acta* **5**, 173 (1972).
5.   E. F. Palermo and J. Chiu, *Thermochim. Acta* **14**, 1 (1976).
6.   A. el Ghany, A. el Kolali and G. Gad, *J. Appl. Chem. Biotechnol.* **21**, 343 (1971).
7.   J. B. Rowse and W. B. Jepson, *J. Thermal Anal.* **4**, 169 (1972).

8. G. S. M. Moore and H. E. Rose, *Nature (London)* **242**, 187 (1973).
9. J. Ruzek, *Silikaty* **18**, 295 (1974).
10. I. Takashima, *J. Jpn Assoc. Min. Petr. Econ. Geol.* **69**, 75 (1974).
11. M. Murat, *J. Thermal Anal.* **3**, 259 (1971).
12. M. D. Judd and M. I. Pope, *J. Appl. Chem. Biotechnol.* **21**, 285 (1971).
13. B. R. Currell and A. J. Williams, *Thermochim. Acta* **9**, 255 (1974).
14. B. Ke, *J. Polymer Sci.* **50**, 79 (1961).
15. B. H. Clampitt, *Anal. Chem.* **35**, 577 (1963).
16. B. Wunderlich, *Thermochim. Acta* **4**, 175 (1972).
17. H. Le Châtelier, *Compt. Rend. hebd. Séanc. Acad. Sci., Paris* **104**, 1517 (1887).
18. J. Orcel and S. Caillère, *Compt. Rend. hebd. Séanc. Acad. Sci., Paris* **197**, 774 (1933).
19. L. G. Berg, I. N. Lepeshov and N. V. Bodaleva, *Dokl. Akad. Nauk. S.S.S.R.* **31**, 577 (1941).
20. R. C. MacKenzie, *Differential Thermal Investigation of Clays*, Mineralogical Society, London, 1957.
21. J. B. Nelson, British Coal Utilisation Research Association, Review 152, of 1955.
22. P. E. Slade and L. T. Jenkins, *Techniques and Methods of Polymer Evaluation*, Vol. I: *Thermal Analysis*, Vol. II: *Thermal Characterisation Techniques*, Arnold, London, 1966 and 1970.
23. B. Ke, *Thermal Analysis of High Polymers*, Interscience, New York, 1965.
24. J. Chiu, *Polymer Characterization by Thermal Methods of Analysis*, Marcel Dekker, New York, 1974.
25. W. Wrasidlo, *Thermal Analysis of Polymers*, Advances in Polymer Science, Vol. 13, Springer Verlag, Berlin, 1974.
26. J. Chiu, *Anal. Chem.* **34**, 1841 (1962).

CHAPTER 10

The Determination of Adsorbed Material on the Surface of Inorganic Solids

This problem is quite widely encountered, particularly in connection with mineral science and technology, as the following examples will demonstrate:

1. The determination of conditioning agents adsorbed onto the surface of mineral particles, in order to effect their separation by froth flotation.
2. The determination of adsorbed flocculating agents, such as polyacrylamide, used in the selective flocculation of suspensions of mineral particles; or, the precipitation of fines from effluent water.
3. The determination of naturally occurring organic material, e.g. fulvic acid, adsorbed on clay minerals.

In most cases, adsorption will have occurred from solution, but DTA can equally well be used to determine organic compounds adsorbed from the vapour phase, providing that the process is not readily reversible. The techniques used for such determinations can be divided into two groups: (a) those which involve solvent extraction of the adsorbed material prior to estimation and (b) those where the adsorbed material is estimated *in situ*. Most of the published methods fall into the first category.

However, it is extremely difficult to obtain complete desorption of the type of compound used as a flotation, or flocculating agent. Very careful control of the experimental conditions is necessary, while the liquids used in solvent extraction are often strongly acidic, or in other ways undesirable. Repeated analysis of the extract is required to ensure that all the previously adsorbed material appears to have been dissolved. Even so, it is very doubtful if chemisorbed material can be removed completely in this way.[1, 2] Another disadvantage, in practice, is that for each type of compound to be estimated, different solvents, extraction procedures and methods of quantitative analysis are likely to be required.

Methods coming into category (b) generally involve either infrared spectroscopy of the adsorbed species, or combustion of the adsorbed film. The difficulties involved in measuring the infrared absorption, or reflectance, spectrum of an adsorbed film, only one molecular layer thick, and covering only part of the surface exposed by the solid particles, are too well known to need further discussion here. In the combustion method, the sample is heated in oxygen and the carbon dioxide liberated is estimated; this, of course, gives no indication of the chemical structure of the adsorbed material or of how strongly it was bound to the surface.

The foregoing discussion has been included to demonstrate the extent of the superiority of the DTA method, which is widely applicable, rapid, generally more accurate and can give additional information on the strength of bonding of the adsorbed material. Basically, the technique involves measuring the heat of combustion of the adsorbed material in a flowing oxygen atmosphere, using a sample of the same chemical composition, but free of adsorbed impurities, as the reference substance.

EXPERIMENTAL PROCEDURE

A detailed account of this technique has already been described elsewhere,[3] as well as its application[4-7] to various mineral/collector systems.

Calibration

Combustion of adsorbed organic material generally occurs within the temperature range 100–450 °C, so that it is first necessary to calibrate the DTA apparatus for calorimetric measurements in this region. The melting of pure indium and of tin, or the two stages of decomposition of $CuSO_4.5H_2O$ can be used for this purpose, the former being preferable.

Next it is desirable to calibrate the apparatus using artificial mixtures of the adsorbate and adsorbent; although if the molar heat of combustion of the adsorbate is accurately known, then the extent of adsorption can be estimated directly from the peak area associated with combustion of the adsorbed film. However, it may then be necessary to apply a correction for the latent heat of vaporization of the adsorbed material, prior to combustion. Mixtures of accurately known composition can usually be prepared by direct mixing of a crystalline organic adsorbate with an inorganic powder. With low melting point organic solids, and with viscous liquids, a convenient method is to prepare a standard solution of the substance in ether. Known volumes of this solution are then added to weighed amounts of the solid adsorbent and the ether allowed to evaporate to constant weight. It is possible that small losses of the adsorbate may occur through a fractional distillation effect, but any errors introduced generally appear to be negligible. The reference material should also be treated in an identical manner with a similar volume of ether, to compensate for the effect of any residues remaining after the evaporation procedure.

In this way, calibration graphs can be prepared relating the extent of adsorption in mg g^{-1} to the DTA exotherm peak area in cm^2, for each of the systems to be studied. An accuracy of the order of 0.1 mg g^{-1} should be obtainable with care, using a Boersma type differential thermal analyser (see Fig. 10.1).

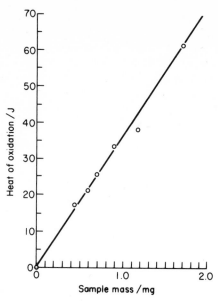

Fig. 10.1. Calibration graph for the combustion of *n*-dodecylamine mixed with titanium dioxide. Ordinate—heat evolved in Joules; abscissa—mass of *n*-dodecylamine in a *c*. 100 mg sample. (From Howe and Pope, Ref. 3.)

Estimation

A weighed sample of the powder carrying an adsorbed film is placed in one DTA cell and a closely similar mass of the same material, known to be free of adsorbed impurities, placed in the reference cell. The sampling procedure is important and a sample mass of about 50–100 mg is recommended. Too small a sample leads to irreproducibility due to lack of homogeneity; too large a sample may lead to slow combustion, limited by the rate of gaseous diffusion, and resulting in broad, irregular exotherms.

An oxygen flow rate of about 500 ml min^{-1} will normally be sufficient to ensure complete and rapid combustion of the adsorbed material, without producing a serious thermal lag, or turbulence in the DTA cell. Nevertheless, the optimum value for a given piece of apparatus can only be determined by experiment. The standard rate of rise of temperature, 10 °C min^{-1}, is quite satisfactory in this work, used in conjunction with a chart speed of about 12 in h^{-1} for recording the temperature. It is highly desirable that the range setting of the differential temperature amplifier should not be changed during a series of measurements, since errors may be introduced which are not easy to correct.

INTERPRETATION

Some typical DTA curves obtained from the combustion of adsorbed reagents on inorganic solids are shown in Fig. 10.2. The area under each exotherm can

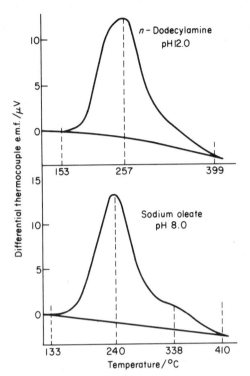

Fig. 10.2. Typical DTA traces obtained from the combustion of absorbed conditioning agents on titanium dioxide. (From Howe and Pope, Ref. 3.)

readily be determined by use of a planimeter, or by the 'cut and weigh' method, as described in Chapter 4. In many cases, the shape of the exotherms change in a progressive manner, as the conditions for adsorption (adsorbate concentration, pH, temperature, etc.) have been varied. The information obtainable from peak shape will obviously depend on many factors and the subject provides considerable scope for further research.

QUALITATIVE APPLICATIONS

In a study of flocculated clay samples, Dollimore et al.[8] found that, although the flocculating agents themselves could not be estimated by DTA, each reagent used had a characteristic effect on the clay dehydroxylation endotherm occurring in the region of 600 °C. Both natural and synthetic polymers were used as flocculating agents. Differential thermal analysis of the flocs was carried out in a nitrogen atmosphere, with a heating rate of 10 °C min^{-1} and using sintered alumina as the reference material. They concluded that it was possible to identify the chemical nature of the flocculating agent by the manner in which the peak and onset temperature of the kaolin dehydroxylation endotherm differed from that of the untreated clay.

The adsorption of fulvic acid on montmorillonite clays has been studied by Kodama and Schnitzer,[9, 10] using DTA at 10 °C min^{-1} in air, again with alumina as reference material. Their results indicated that it was possible to distinguish between fulvic acid adsorbed on the external surface of clay particles and inter-lamellar adsorption. Externally adsorbed fulvic acid gave rise to an exotherm on the DTA curve within the approximate temperature range 300–500 °C, whereas for interlamellar adsorption the exotherm occurred between about 500 and 800 °C.

Differential thermal analysis has also been employed by Eltantawy and Arnold,[11, 12] in qualitative studies of the adsorption of various organic compounds on clay minerals.

ADSORPTION OF WATER

Various attempts have been made to use DTA to investigate the extent of water adsorption[13] and the strength of bonding of adsorbed water,[14] so as to identify the existence of various types of surface site. However, such applications appear to be highly specialised and usually require evidence from other techniques. Differential thermal analysis cannot at present be considered as being generally applicable to the determination of water vapour adsorption isotherms. However, it may well be possible to distinguish between physical and chemisorbed water from the temperature range over which desorption occurs on heating.

REFERENCES

1. S. J. Gregg, *The Surface Chemistry of Solids*, 2nd Edn, Chapman and Hall, London 1965.
2. S. I. Pol'kin, S. F. Laptev *et al.*, Proceedings of the Xth International Mineral Processing Congress, London, 1973, Paper 35.
3. T. M. Howe and M. I. Pope, *Powder Technol.* **4**, 338 (1971).
4. M. I. Pope and D. I. Sutton, *Powder Technol.* **5**, 101 (1972).
5. M. I. Pope and D. I. Sutton, *Powder Technol.* **7**, 271 (1973).
6. M. I. Pope and D. I. Sutton, *Powder Technol.* **9**, 273 (1974).
7. M. I. Pope and D. I. Sutton, *Powder Technol.* **10**, 251 (1974).
8. D. Dollimore, G. R. Heal and T. A. Horridge, *Clay Minerals* **8**, 479 (1970).
9. H. Kodama and M. Schnitzer, *Proceedings of the International Clay Conference, Tokyo*, Vol. 1, 1969, p. 765.
10. H. Kodama and M. Schnitzer, *Can. J. Soil Sci.* **51**, 509 (1971).
11. I. M. Eltantawy and P. W. Arnold, *Nature (London) Phys. Sci.* **237**, 123 (1974).
12. I. M. Eltantawy and P. W. Arnold, *J. Soil Sci.* **24**, 232 (1973).
13. E. Berlin, P. G. Kliman and M. J. Pallansch, *Thermochim. Acta* **4**, 11 (1972).
14. V. V. Morariu and R. Mills, *Z. Physik Chem. Neue Folge* **78**, 298 (1972).

CHAPTER 11

Studies of Catalytic Activity and the Rates of Heterogeneous Reactions

Much of the published work involving the use of DTA in studies of catalysts has been confined to establishing the phase composition of the catalyst, or to catalyst identification. The method of determination of phase diagrams for condensed systems has already been discussed in an earlier section; its application to catalysts is exactly the same as for any other compounds of similar chemical composition. Nevertheless, phase composition has been shown to be an extremely important factor in determining both catalytic activity and selectivity. This is well illustrated by the case of bismuth molybdate catalysts[1-3], which are widely used for the partial oxidation and ammonoxidation of hydrocarbons to form monomers for the plastics industry. The application of DTA to the study of single, binary and ternary oxide catalyst systems has been reviewed by Bhattacharyya and Datta.[4]

Another field of application has been to reactions between solids used in the preparation of catalysts. Again the technique has been described in a preceding section.

The use of DTA to evaluate the ability of materials to act as catalysts for reactions normally occurring in the gas phase is a somewhat more complex subject.

Although the overall catalytic activity can be determined fairly readily, DTA will not provide any evidence as to the selectivity (product distribution). For example, in the catalytic partial oxidation of hydrocarbons, it is essential to know how much of the reactant has been converted to the corresponding olefine and what proportion is wasted by the formation of such products as carbon monoxide, carbon dioxide and water. To obtain this information, it is necessary to analyse the effluent gases from the DTA cell (where a flow system is used) e.g. by gas–liquid chromatography.

The main advantages of the DTA method are, firstly, that measurements can be made on a large number of materials in a relatively short time, so that the choice of compounds for a detailed investigation can be narrowed down considerably.[5] Second, if the gaseous reactants are passed at a controlled rate through a packed bed of catalyst in a DTA cell, the temperature of which is known, it is possible to simulate the conditions likely to exist in a commercial catalytic reactor. However, some caution must be exercised in interpretation of such data, because of the large temperature gradients that inevitably exist in the massive catalyst bed of a commercial reactor. This is due to the heat evolved by the reaction; as the concentration of reactants declines during passage through the catalyst bed, so does the corresponding rate of heat evolution.

Measurements of catalytic activity can be made in both static and flowing atmospheres; but the latter procedure, although more complicated, provides information of far greater value.

STATIC SYSTEMS

In this method, the catalyst is placed in the DTA sample container, while the reference cell is either left empty, or filled with a material known to be catalytically inert. Outgassing is effected by heating the catalyst in a flowing atmosphere of an inert gas; helium, argon, or in some cases nitrogen, may prove suitable for this purpose.

After cooling in the inert atmosphere, measured amounts of the reactants are introduced into the gas stream. The apparatus is thoroughly flushed out with this mixture of accurately known composition and then closed. The DTA head is now heated at a programmed rate; the temperature at which the reaction is first detected can be used as a measure of catalyst activity.

However, the conditions used bear little comparison with those found in a commercial plant. At best, the information obtained is of qualitative use only.

FLOW SYSTEMS

The main problem encountered with flow systems is that it is still generally necessary to construct one's own measuring head, for use on the DTA apparatus. Certainly, DTA heads recommended for use with catalysts are marketed by various manufacturers, but most suffer from limitations which restrict their application. Accordingly numerous papers have recently appeared describing the construction of measuring heads, or the modification of those available commercially.

Apparatus

The most important consideration is that the reactant gases must pass through, and not over, the packed bed of catalyst. For many reactions, it is also necessary to ensure that the reactant gases do not come into contact with the thermocouple wires, because of the well-known catalytic properties of noble metals. The use of platinum sample containers is obviously unacceptable, while many other metals are also open to objection. An apparatus which overcomes these problems has recently been described by Schubert and Barth,[6] although it is unsuitable for use at high temperatures, because of its glass construction. However, there appears to be no reason why a similar apparatus could not be made from silica or a ceramic material. Details of the measuring head are illustrated

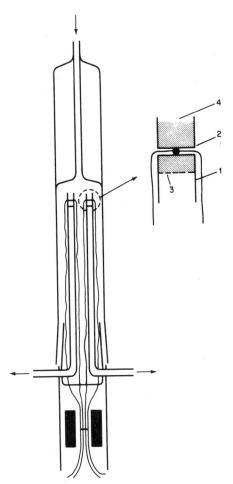

Fig. 11.1. Differential thermal analysis head designed for catalyst studies (from Schubert and Barth, Ref. 6.) The cell (1) is made of pyrex glass and is 6 mm in diameter; the horizontal tube (2) supports the thermocouple junction, which is thus protected from exposure to the reactant gases. A porous plate (3) supports the bed of catalyst. (4) through which the reactant gases are passed.

in Fig. 11.1. The catalyst is contained in a glass cell of 6 mm diameter and is supported by a glass plate perforated with holes of approximately 0.1 mm diameter. A horizontal glass tube passes through the centre of the cell and supports the thermocouple junction, which is thus protected from contact with the reactant gases passing through the bed of catalyst.

Experimental

As with the static system, the catalyst powder is placed in the DTA measuring cell and the reference cell either left empty, or filled with an inert material. Some workers prefer to prevent gas-flow through the reference cell, but the advantages obtained by this procedure are unclear. Outgassing of the catalyst involves heating in a flowing atmosphere of an inert gas, until no further desorbed material can be detected. Since catalytic studies are nearly always carried out in conjunction with some method of evolved gas analysis (EGA) for product estimation, this presents no problem.

Subsequent measurements of catalytic activity can either be carried out with the catalyst maintained at a constant temperature, or using a controlled rate of rise of temperature. In both cases, the flowing atmosphere containing the reactants should be preheated to the temperature of the catalyst.

Various methods are available for introducing the reactants into the stream of inert carrier gas. With volatile organic liquids, the carrier gas may simply be bubbled through a sinter immersed in the liquid, but it is difficult to obtain adequate control of the vapour concentration in this way. Alternatively a known volume of liquid, or gaseous reactants, can be injected into the carrier gas stream with a hypodermic syringe and septum cap, following the procedure widely used in gas chromatography. With gaseous reactants, controlled proportions can be fed continuously into the carrier gas stream via flow meters and needle valves. Yet another method is to use a gas switching valve, where known volumes of gaseous reactants, of previous prepared composition, are injected into the carrier gas stream. This has several advantages over the use of a hypodermic syringe, e.g. a gas switching valve is generally more accurate and avoids the risk of introducing air into the system.

Interpretation of results

Where a single charge of the reactants is injected into the carrier gas stream, with the catalyst maintained at a pre-determined temperature, adsorption and subsequent reaction on the catalyst surface leads to an exothermic peak. The area under this peak for a given dose of reactants can be taken as a measure of the effectiveness of the catalyst at that particular temperature, providing that the reaction has not gone to completion.

If a known, constant, partial pressure of reactants is introduced into the carrier gas stream, then exothermic reaction at the catalyst surface will normally lead to a steady temperature differential being set up between the catalyst and the reference cell. The height of the DTA trace above the base line then gives a measure of catalytic activity, at a given temperature and reactant pressure. This latter procedure most closely reproduces the conditions encountered in an industrial process. Failure to establish a steady temperature differential may be due to poisoning of the catalyst, or the blocking of surface sites by strongly adsorbed products. Where blocking has occurred, an increase in catalyst temperature will probably lead to more rapid desorption of products and hence to the establishment of a steady state.

Use of a programmed rate of rise of temperature, in conjunction with a constant flow rate of reactants over the catalyst, leads to deflection of the DTA trace at the lowest temperature at which the reaction becomes detectable. This temperature gives an indication of the relative activity of the catalyst and of the lowest temperature at which the system can be studied in detail.

APPLICATIONS

The main uses of DTA in catalyst evaluation are as follows:

(a) Studying the effectiveness of a wide range of different substances for catalysing a given chemical reaction. This provides a quick and simple means of selecting those compounds worthy of a detailed investigation.

(b) Studying how different methods of catalyst preparation and support influence the activity of any given substance, or combination of substances.

(c) Measuring the ageing, or deactivation of catalysts after a period of service. The purpose of such measurements may be either to decide if catalyst replacement has become necessary, or to investigate methods of prolonging catalyst life.

(d) Discovering what substances act as poisons for a given catalyst, by injecting small traces into the reactant gas stream.[10]

CATALYST BEAD SYSTEMS

A technique for measuring the rates of heterogeneous reactions using catalyst beads deposited on thermocouple junctions, or platinum heating coils, has been described by Firth.[7] Where thermocouple junctions are used, the apparatus is heated by an external furnace, as in conventional DTA. The platinum coils, on the other hand, serve both as heaters and as resistance thermometers. With both systems, it is important that the catalyst beads should have a very small heat capacity and that the reaction to be studied should be accompanied by a fairly large evolution of heat. The technique has been developed for the quantitative study of the rates of heterogeneous catalytic reactions by Cullis, Nevell and Trimm.[8]

The reactor beads can be prepared by coating the small platinum coils (or thermocouple junctions) with alumina, by dipping them into a concentrated solution of aluminium nitrate, followed by heat treatment. A metallic deposit, which acts as the catalyst, is then obtained by dipping one of the beads into a solution containing the salt of a noble metal, with subsequent further heat treatment. Where a platinum coil is used, the catalyst bead and the inert reference bead are connected into a d.c. bridge circuit; the bridge current is adjusted so as to heat the beads to the desired reaction temperature. A charge of the reactant gases is then introduced to the system and the rate of the catalysed reaction followed by measuring the temperature difference set up between the catalyst and reference beads; the platinum coils are now serving as resistance thermometers.

For kinetic measurements, it must be remembered that the technique inherently involves non-isothermal conditions. However, the measurements can be corrected to correspond to the temperature at the end of the reaction, if the temperature coefficient of resistance for the catalyst bead and an approximate value for the energy of activation of the reaction are known. A detailed description of the method and underlying theory of the kinetic calculations can be obtained from papers by Cullis et al.,[8] and by Jones et al.[9]

The catalyst bead system was originally developed for detecting the presence of methane, or other potentially explosive gas–air mixtures, in industrial atmospheres.

A sample of contaminated air can be collected and a known volume subsequently introduced into the flowing atmosphere of a DTA apparatus consisting of a catalyst, and inert reference bead. The heat liberated by combustion of the sample then gives a measure of the amount of contamination present. In this way, DTA can be used to monitor the atmosphere in coal mines or in potentially hazardous areas of petrochemical and polymer plants.

PULSED THERMOKINETIC (PTK) MEASUREMENTS

A technique has been introduced by Richardson et al.[11] which allows study of both adsorption and reaction of gases on the surface of catalysts in a packed bed. The apparatus consists essentially of a differential thermal analyser connected to a katharometer for evolved gas detection; alternatively the evolved gases can be switched to pass through a chromatographic column, before arriving at the katharometer. Helium was used as the carrier gas, with samples of reactant being introduced into the gas stream by means of a pulsing valve. In this way the adsorption of hydrogen on a nickel catalyst has been shown to give a sharp exothermic peak on adsorption, followed by a small, broader endotherm, corresponding to desorption of some of the gas. As further pulses of hydrogen are admitted to the carrier gas, so the catalyst surface becomes saturated with chemisorbed hydrogen. Eventually the adsorption exotherm and subsequent desorption endotherm become identical, corresponding to reversible adsorption; the peak area then gives a measure of the amount of reversible adsorption that has occurred.

Similar measurements have been made[12] with ethylene and ethane, enabling both heats of adsorption and the kinetics of ethylene hydrogenation to be determined.

This technique seems to be applicable to a wide range of catalysed reactions; both argon and hydrogen have been used as carrier gases by Richardson et al.

REFERENCES

1.   Ph. A. Batist et al., J. Catal. 12, 45 (1968).
2.   P. Boutry, R. Montarnal and J. Wrzyszcz, J. Catal. 13, 75 (1969).

3. B. Grzybowska, J. Haber and J. Komorek, *J. Catal.* **25**, 25 (1972).
4. S. K. Bhattacharyya and N. C. Datta, *J. Thermal Anal.* **1**, 75 (1969).
5. K. Papadatos and K. A. Shelstad, *J. Catal.* **28**, 116 (1973).
6. J. Schubert and F. Barth, *Chem. Tech. Berlin* **25**, 163 (1973).
7. J. G. Firth, *Trans. Faraday Soc.* **52**, 2566 (1966).
8. C. F. Cullis, T. G. Nevell and D. L. Trimm, *J. Physics E* **6**, 384 (1973).
9. A. Jones, J. G. Firth and T. A. Jones, *J. Physics E* **8**, 37 (1975).
10. R. J. Farrauto and B. Wedding, *J. Catal.* **33**, 249 (1974).
11. J. T. Richardson, H. Friedrich and R. N. McGill, *J. Catal.* **37**, 1 (1975).
12. J. T. Richardson and H. Friedrich, *J. Catal.* **37**, 8 (1975).

CHAPTER 12

Liquid Crystals

Most crystalline solids, when heated, eventually melt at a well defined tempera-ture to form a mobile, isotropic liquid. Use of DTA in determining the tempera-ture and latent heat of fusion of such materials has been discussed in a previous chapter. However there exists a range of organic compounds which do not behave in this manner. Instead these crystalline solids first melt to form a turbid fluid, which only transforms to a clear, isotropic liquid at a significantly higher temperature. This phenomenon was first discovered by Reinitzer in 1888 and appeared to be associated[1] with organic compounds having a molecular struc-ture consisting of elongated molecules, giving rise to a certain amount of struc-tural rigidity. Examples of the types of compound which behave in this manner include: cholesteryl esters, sodium soaps of aliphatic acids, *p'p'n*-alkoxyazo-benzenes, anisaldazine and even some unsaturated aliphatic carboxylic acids.

The turbid, fluid phase formed by melting of the crystalline solid is said to contain liquid crystals, of which several distinct types have been identified. *Smectic* liquid crystals are the first to be formed; the molecules normally orientate themselves so that their long axes lie parallel to one another, forming layers. These layers are able to slide, or rotate relative to each other, thereby allowing the phase a certain degree of mobility. With increasing temperature, the liquid crystals may pass through several distinct smectic phases, which differ from each other in their degree of two-dimensional ordering. Eventually the smectic phase either melts to an isotropic liquid, or it may transform to a *Nematic* phase at a characteristic temperature. In a nematic phase, the molecules still lie with their major axes parallel to one another, but there are no distinct layers; hence the degree of long range order is lower and the fluid is more mobile.

Where the individual molecules are optically active, a *Cholesteric* rather than a nematic phase may be formed, having a helical type of structure.

Some compounds form only a smectic phase, while others give only a nematic, or cholesteric phase. When liquid crystals form both on heating the crystalline solid and on cooling the isotropic liquid, they are called enantiotropic. In some cases, the liquid crystals are only observed if the isotropic liquid is cooled below the melting point of the crystalline solid; they are then called monotropic. An example[2] of this type of behaviour is given by cholesteryl laurate, which has a normal melting point of 99.0 °C, but which on cooling forms liquid crystals having transition temperatures of 87.4 and 80.7 °C.

Considerable interest has recently been shown in the study of liquid crystal transitions, either because of their importance in the development of new electro-optical devices, or because of their role in lipid–water systems. While some transitions can be observed directly by hot stage microscopy, others have so far escaped detection. In any case, a great deal depends on the skill and judgment of the observer.

The advantages of using DTA, preferably in conjunction with hot stage microscopy, are overwhelming. Probably the main difficulty arises through the low magnitude of the transition enthalpy changes involved, which often have values of less than 0.2 J g^{-1}. However, such measurements are well within the capability of DTA and DSC apparatus now available, as demonstrated by the work of Barrall,[2] Gray[3] and others.

For a comprehensive treatise on the physics and chemistry of liquid crystals, the reader is advised to consult Refs 1 and 8. A large amount of data on the thermodynamics of liquid crystal transformations have been obtained and correlated by Barrall et al.[2, 4, 7] mainly through use of DTA.

MAGNITUDE OF THE ENTHALPY CHANGES ASSOCIATED WITH LIQUID CRYSTAL TRANSITIONS

From the data listed by Barrall et al.[2, 4, 7] it is clear that the enthalpy of melting of the crystalline solid is always substantially greater than that of subsequent liquid crystal transitions. Gray[5] suggests a range of values indicated below.

Transition	$\Delta H/kJ \ mol^{-1}$
Melting of crystalline solid	20–170
Smectic → nematic or cholesteric Smectic → isotropic liquid	4–20
Smectic → smectic	0.5–10
Nematic → isotropic liquid Cholesteric → isotropic liquid	< 0.5 – 3.5

The magnitude of the enthalpy and entropy changes accompanying liquid crystal transitions provides useful information on the structure and degree of ordering in the various phases.

EFFECT OF PRESSURE ON LIQUID CRYSTAL TRANSITIONS

Because of the relatively small energy differences that exist between certain liquid crystal states, the transition temperatures are sometimes very much influenced by changes in pressure. This has been investigated by Garn and Richardson,[6] who found that for a series of bis(4'-n-alkoxybenzal)-1,4-phenylenediamines, the temperatures of successive smectic–smectic and the smectic–nematic transitions increased by up to 4 °C on raising the pressure from one to ten atmospheres.

EXPERIMENTAL PROCEDURE

Because of the large number of consecutive phase transitions that may occur when the crystalline solid is heated to a temperature above its melting point, and the small physical differences between these phases, the corresponding DTA curve is very difficult to interpret without additional information. It is therefore highly desirable to use a DTA apparatus which is so designed that the test sample can be observed directly, using a polarizing optical microscope. Several firms now manufacture micro-DTA apparatus fulfilling this requirement and being capable of controlled cooling as well as controlled heating. The importance of controlled cooling is due to the fact that some liquid crystal phases are only observed during the cooling cycle, as discussed previously.

It is necessary to use a small sample, e.g. <10 mg, in order to obtain good peak resolution, while the apparatus must be sufficiently sensitive to record transitions where the heat change involved may often be less than 1 J g^{-1}. Clearly, then, DTA equipment required for liquid crystal studies must be chosen carefully. A relatively slow heating rate is desirable and many workers seem to have selected 5 °C min^{-1} as being most satisfactory. As with normal melting measurements, a static, inert atmosphere should be used, but it will not of course be possible to encapsulate the sample and still observe it during the experiment.

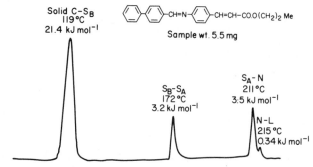

Fig. 12.1. DTA curve for a compound showing smectic–smectic and smectic–nematic transitions. (From Gray, Ref. 5.)

An example of the kind of DTA curve which can be obtained is illustrated in Fig. 12.1, taken from the work of Gray.[5] These measurements were made using a Standata 671B DTA apparatus, at a heating rate of 5 °C min^{-1} in a static nitrogen atmosphere. It will readily be seen that melting of the crystalline solid (C) to form the low temperature smectic phase (S_B) involves the largest enthalpy change, while melting of the nematic phase (N) to give an isotropic liquid (L) is associated with the smallest.

REFERENCES

1. G. W. Gray, *Molecular Structure and the Properties of Liquid Crystals*, Academic Press, New York, 1962.
2. E. M. Barrall II, J. F. Johnson and R. S. Porter, *Proceedings of the Second International Conference for Thermal Analysis*, Vol. 1, (R. F. Schwenker and P. D. Garn, Eds), Academic Press, New York, 1969, pp. 555–570.
3. G. W. Gray, *Some Applications of DTA to Studies of Liquid Crystalline Systems*, Meeting of the Thermal Methods Group of the Society for Analytical Chemistry, London, 1971.
4. R. S. Porter, E. M. Barrall II and J. F. Johnson, *Proceedings of the Second International Conference for Thermal Analysis*, Vol. 1, (R. F. Schwenker and P. D. Garn, Eds), Academic Press, New York, 1969, pp. 597–613.
5. G. W. Gray, *Differential Thermal Analysis of Liquid Crystal Transitions*, Technical Information Sheet No. 14, Stanton Redcroft, London.
6. P. D. Garn and R. J. Richardson, *Proceedings of the Third International Conference for Thermal Analysis*, Vol. 3, Berkhauser Verlag, Basel, 1972, pp. 123–130.
7. E. M. Barrall II and J. F. Johnson, *Liquid Crystals and Plastic Crystals*, Vol. 2 (G. W. Gray and P. A. Winsor, Eds), Ellis Horwood, Chichester, 1974, pp. 254–306.
8. E. B. Priestley, P. J. Wojtowicz and P. Sheng, *Introduction to Liquid Crystals*, Plenum Press, New York, 1975.

CHAPTER 13

Ferroelectric Transitions—the Curie Point

In this discussion, the Curie point will be defined as the temperature at which a crystalline solid undergoes a transition from a ferroelectric to a paraelectric phase. Methods which have been used to determine this temperature include measurement of dielectric constant, magnetic properties, dilatometry, neutron diffraction and Mossbauer spectroscopy. More recently, the improved sensitivity and reproducibility of commercially available differential thermal analysers, has made it possible to determine Curie points by DTA.

Although some ferroelectric transitions appear to be accompanied only by a change in heat capacity (i.e. they are second order), in the majority of cases studied an appreciable enthalpy change is involved. Such transitions therefore give rise to a small peak on the DTA curve.

In fact DTA has several advantages over the conventional methods of determining Curie points. Only relatively small samples weighing, less than 10 mg are necessary and these do not need to be in the form of single crystals. Furthermore, problems associated with attaching conducting electrodes to the crystal are eliminated, together with subsequent difficulties resulting from electrical 'noise'. Another advantage is that the crystalline sample does not have to be orientated relative to the polarization axis.

EXPERIMENTAL PROCEDURE

Because the total enthalpy changes associated with ferroelectric transitions are comparatively small, sometimes as low as 5 J mol^{-1}, a DTA apparatus capable of high sensitivity with small samples will be necessary. Controlled cooling as well as controlled heating will often be required, for two reasons. Firstly, many ferroelectric transitions occur at subambient temperatures, e.g. KH_2PO_4 at 122 K, so that an apparatus cooled from a liquid nitrogen source is to be recommended. Secondly, Curie points frequently exhibit temperature hysteresis; a detailed study of this effect requires consecutive heating and cooling cycles at a slow and accurately controlled rate.

Since the measured enthalpy changes may well be very small, it is particularly important to match both the thermal conductivity and heat capacity of the test and reference sample, thereby minimizing base-line drift. Nevertheless, heat-treated, high purity alumina may often prove suitable as a reference substance.

The heating rate is also fairly important, since too slow a rate will give rise to broad, indistinct peaks; conversely, too high a rate will accentuate the effects of hysteresis. Probably a heating rate of 5 °C min^{-1} will prove most satisfactory in the majority of cases. The ambient atmosphere has little influence on ferro-electric transitions, so that static air can generally be used.

INTERPRETATION

The problems associated with interpretation of DTA peaks resulting from solid–solid phase transitions have been discussed in Chapter 4. As yet it is unclear whether peak, or extrapolated onset temperatures give the most satisfactory values of Curie points. Loiacono[1] has demonstrated remarkably good agreement between Curie points obtained from DTA curves, using the extrapolated onset temperature, and accepted literature values. These data, taken from his paper, are recorded in Table 13.1. However, Abell et al. in a detailed study[2] of barium

TABLE 13.1

Some Curie temperatures (T_c) of representative ferroelectric crystals, as determined from the extrapolated onset temperature of DTA peaks, compared with corresponding literature values (from Loiacono Ref. 1).

Crystal	Present work T_c/K	Literature T_c/K
KH_2PO_4	121.8	121.97
RbH_2PO_4	142.1	147.1
$NH_4H_2PO_4$	148.5	149
CsH_2PO_4	149.3	159
$NH_4H_2AsO_4$	219.2	220
CsH_2AsO_4	139.2	143.3
KD_2PO_4	226.3	220.6
RbD_2PO_4	227.2	226.4
$ND_4D_2PO_4$	237.9	235.1
CsD_2AsO_4	209.4	212.4
$BaTiO_3$ (a)	398.2	393
(b)	284.9	300
(c)	180.6	193
$LiTaO_3$	903.7	891
$Ba_2NaNb_5O_{15}$	853.5	833
$Gd_2(MoO_4)_3$	429.5	432
TGS	323.2	322.1
TGFB	347.6	346
$SC(NH_2)_2$ (a)	167.9	169
thiourea (b)	200.6	180
$NaKC_4H_4O_6 . 4H_2O$ (a)	not detected	297
Rochelle salt (b)	not detected	255
$(NH_4)_2SO_4$	225.7	223.7

sodium niobate crystals, found that while peak temperatures agreed quite well when comparing heating and cooling curves, the extrapolated onset temperatures differed considerably. This difference may of course be a genuine effect due to hysteresis in the phase transition. Their results are set out in Table 13.2

TABLE 13.2

Curie temperatures measured by DTA on various compositions within the B-S-N phase field.[2]

Composition/mol %			Curie temperature T_c/°C			
			Heating		Cooling	
Na_2O	BaO	Nb_2O_5	Onset	Peak	Onset	Peak
1 7.3	41.6	51.1	552	562	566	562
2 7.8	40.5	51.7	552	560	568	560
3 8.0	41.0	51.0	560	568	575	570
4 8.2	40.7	51.1	572	580	585	580
5 8.3	40.1	51.6	575	580	583	578
6 8.3	38.7	53.0	565	575	582	575

for crystals of varying composition. It will be seen that using a heating rate of 5 °C min⁻¹, a difference of up to 10 °C was observed between the extrapolated onset and the peak temperature. A typical DTA curve obtained by these workers is shown in Fig. 13.1.

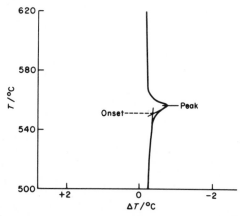

Fig. 13.1. Typical DTA trace in the vicinity of the Curie point of B-S-N. The construction shows how one measurement of T_c was taken; the other was taken as the peak temperature. (From Abell *et al.*, Ref. 2.)

REFERENCES

1.   G. M. Loiacono, *Ferroelectrics* **5**, 101 (1973).
2.   J. S. Abell, I. R. Harris and B. Cockayne, *J. Mater. Sci.* **8**, 667 (1973).

CHAPTER 14

Polymers

Differential thermal analysis studies on polymeric materials are numerous and many books and reviews have been published on the subject[1-6]. It is impractical to attempt to summarize all the work that has been published on this subject and therefore this chapter gives a more generalized description of the use of DTA in the study of polymers. Specific cases are described where this is necessary to illustrate the overall subject, and references are given to the relevant literature. For detailed descriptions the Reader is advised to consult the references given above.

GENERAL CONSIDERATIONS

An immediate problem which confronts the experimenter is that the materials to be studied are frequently not in a form which enables easy handling. There is obviously no problem when the polymer can be obtained in powdered form (other than those effects discussed in Chapter 3) since the sample can be treated in a manner analogous to that used for an inorganic salt. However, it is an unfortunate fact of life that the vast majority of polymers cannot be obtained in this ideal sample geometry.

For thermoplastic materials it has been suggested that to obtain the most efficient sample packing, the material should first be melted in the sample container.[7] This will ensure that good contact is made between the sample and the walls of the container. With such a procedure, however, caution has to be exercised in ensuring that the polymer is stable up to and just beyond its melting point, with respect to both chemical and physical properties.

If the sample is received for investigation in the form of a large block, then the first problem is the choice of a suitable method by which this can be reduced to a size compatible with the sample crucible. In the case of materials with a soft or rubbery consistency, such as polyurethanes, silicones, etc. then it is an easy process to cut small pieces off the block and to pack these carefully into the crucible. With the more rigid materials, such as epoxies, then other methods have to be used. With inorganic materials the process of grinding is frequently carried out in order to reduce agglomerates down to a smaller particle size. However for polymer studies this is not a recommended procedure, since the heat generated during the grinding process may well cause partial degradation of the sample.

A good example of such an effect was once seen by the authors when asked to examine a sample of a polymer used in the construction of a superconducting magnet. Samples were supplied in the form of a coarse powder and the information requested was the measurement of any glass transition temperature in the range 0–100 °C. On heating the sample (in air) a multitude of small endothermic peaks were observed on the DTA trace, rendering the observation of any glass transition temperature hopeless. When the sample was cooled and reheated these peaks had disappeared and the glass transition was readily apparent. After details of the sample history were requested, it was found that the powdered samples had been prepared by the drilling of a large block of material. Undoubtedly the heat generated by the drilling process had caused some reaction to occur and this was reflected by the peaks on the DTA trace. (A subsequent TG experiment showed that the initial heating from 0–100 °C was associated with a 1 % weight loss.) A possible way of overcoming this problem is to grind the material at a low temperature in order to reduce the heat effects.[8]

The choice of a suitable reference material may well prove difficult in that the normal inorganic compounds used (such as alumina and silica) have considerably different heat capacities and thermal conductivities compared to most polymers. This is alleviated somewhat in the case of filled polymers, which may well contain large amounts of an inorganic material. Where low sample weights are used for the DTA experiment, it may well be found advantageous not to use a reference material at all and just utilize an empty crucible. When larger sample weights are used it will probably prove necessary to investigate two or more different reference materials (which satisfy the criterion of inertness) before deciding on which to employ. Silicone oils are frequently used as a reference because their properties closely match those of polymers. Because of the presence of large numbers of carbon atoms in a polymer chain it is generally not considered advisable to use a diluted sample, due to the possibility of reaction between sample and diluent.[9]

GENERAL DESCRIPTION OF THE DATA WHICH CAN BE OBTAINED BY DTA

Three processes are particularly suited to observation by DTA. These are:

(a) glass transitions;
(b) melting and associated crystallization effects;
(c) degradation.

An idealized DTA trace in air is shown in Fig. 14.1 and consists of the above-mentioned effects. The exothermic oxidation peak assumes that the heat associated with the oxidation of any carbon, or carbon monoxide, formed outweighs the absorption of heat by the endothermic decomposition process. In general, under an inert atmosphere, this last peak will be endothermic, although polymers where the decomposition process is exothermic are not unknown.

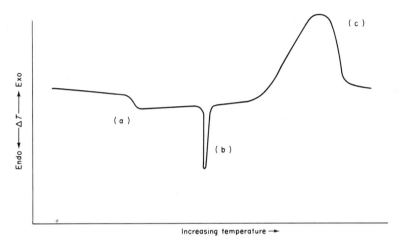

Fig. 14.1. Hypothetical DTA trace for a polymer heated in air.

Glass transition temperature, T_g

An exact definition of the 'glass transition temperature' is difficult to give. If we consider an amorphous linear polymer it may have a structure which is relatively hard and inelastic, or it may have the converse, i.e. be highly elastic like a rubber. The glass transition temperature may be defined as that temperature below which the material loses this elastic character. However, the change in state of the polymer at T_g involves no major changes in the basic physical properties of the material and is only associated with changes in specific heat, thermal expansion coefficient and free volume. Since no change in latent heat is involved, this phenomenon is classed as a second-order transition and is usually manifested in DTA as a drastic change in baseline. Figure 14.2 illustrates the general manner in which a glass transition is exhibited by DTA. Keavney and Eberlin[10] determined the glass transition temperatures for a wide range of polymeric materials. They showed that for polystyrene and polyvinylchloride the glass transition was associated with a ΔT value of 0.1–0.2 °C, whereas a first-order transition for a comparably sized sample would be 10–50 times as great. Therefore it is apparent that in order to measure the glass transition using DTA, the equipment must possess high sensitivity and good baseline stability.

The transition temperature is not well defined, but varies with the heating (or cooling) rate due to the time dependence of the relaxation of long chain molecules. Thus a high heating rate may well yield a high value for T_g whereas measurements made at slow heating rates may yield no value at all, due to the change in specific heat occurring over a long period of time. Such effects have been reported by

Schwenker and Zuccarello[11] and Keavney and Eberlin;[10] these latter workers, however, found that this variation was minimal at heating rates in the range 1–6 °C min^{-1}.

In many instances, as illustrated in Fig. 14.2, it is difficult to decide as to the exact temperature to take as referring to T_g. The generally accepted way of

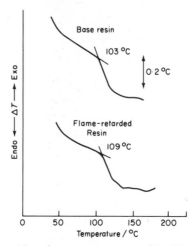

Fig. 14.2. Glass transitions for two prepreg materials. (From Blaine, Ref. 31.)

obtaining the temperature is by using the method outlined in Chapter 5. Temperatures obtained by this method are the most widely reported in the literature and therefore this approach is to be preferred, if only by reason of it giving data comparable to those already published.

Many studies have been published on the use of DTA in the determination of T_g (e.g. Refs 12–14); for a detailed discussion of the glass transition itself the reader should consult the review published by Kovacs.[15]

Melting

The determination of the melting point of a polymeric material can readily be made by DTA. As indicated by Fig. 14.1, the effect is represented as a sharp endotherm on the DTA trace. However, because of other processes which may occur in the same temperature range, it is often necessary to confirm the melting temperature by visual observation of the sample. This can be done using conventional hot stage microscopy or, more preferably, by visual observation of the DTA sample. Conventional low temperature DTA equipment often has a glass window built into the top of the furnace assembly. By mounting a suitable microscope with top illumination above the furnace the sample in the DTA cell can be observed continually during the DTA experiment. An example of another form of equipment suitable for simultaneous DTA and visual observation is described in Chapter 2.

The melting point of a polymer is very dependent on its history and particularly on the degree of crystallinity within the sample.[17-19] In fact this has been used as a measure of the degree of crystallinity within a sample of polyethylene.[20] In this method a sample is heated rapidly to above its melting point, cooled under controlled conditions and reheated rapidly. The melting point then will be lower than that for the well annealed material (138.5 °C); this difference is a measure of the crystallinity within the sample. (See also p. 82.)

The size and shape of any crystallites present in the polymer sample will affect the broadness of the melting endotherm, because smaller crystallites will melt at a lower temperature than the larger crystallites.

In addition to the crystallinity already present in the sample, further crystallization may be induced into the sample by heating. This is observed on the DTA trace as a sharp exotherm proceeding the melting endotherm. Several reports have been published which demonstrate this effect.[11,21,22]

When determining melting points for polymers it may be important to eliminate this possibility of including additional crystallinity, since this may well affect the value of T_m obtained. This can be accomplished by using heating rates which are sufficiently fast to prevent reorganization within the material. Scott[23] was able to define the melting point of polyethylene terephthalate to within ± 0.5 °C (260 °C). Ke[24] has shown the effect of a diluent (phenanthrene) on the melting point of polyethylene and his results indicate that there is a marked depression of the melting point as the concentration of phenanthrene is increased.

Decomposition

The decomposition of a polymeric material, under an inert atmosphere, is generally an endothermic reaction. In an atmosphere of normal air the process is generally complicated by the superimposition of oxidative reactions. These latter effects usually make a higher heat contribution to the overall mechanism and are therefore predominant on the DTA trace. Since these processes are affected by changes in sample weight and heating rate, the precautions outlined in Chapter 3 should be followed. In the case of epoxy resins, it is frequently found that in an inert atmosphere there is a pronounced exothermic contribution due to the isomerism or polymerization of residual epoxy groups.[25]

A point which is generally applicable to polymer decomposition reactions is that DTA is of little use when used as the only analytical tool. It is practically impossible to interpret the DTA trace without some additional information. This can be obtained by the use of two other techniques, namely thermogravimetry and effluent gas analysis. The use of mass spectroscopic techniques to study decomposition reactions has received widespread interest in recent years,[26,27] although most of this has been restricted to inorganic materials. There is a very good reason for this; a polymer degradation process is an extremely complicated reaction involving the formation of many different products. Many of these products are capable of condensation and thus never reach the mass spectrometer source. Others reach the mass spectrometer, but then proceed to repolymerize under the high ionization voltages which are applied, thus giving a high contamination level. The ideal answer is to utilize the combined advantages offered by a gas chromatograph–mass spectrometer system, but this represents a considerable financial investment.

Many studies have been reported on the decomposition of polymeric materials and it is impossible to summarize these here. The reader should consult the references given at the end of this chapter for more detailed information. However it is of use to consider a few examples in detail in order to show how DTA may be used.

Kotoyori[28] has studied the thermal degradation of several commercial plastics in an atmosphere of air. An example of the results obtained is given in Fig. 14.3

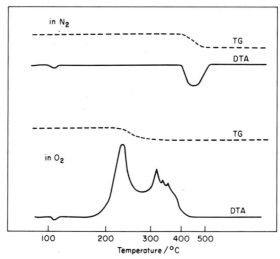

Fig. 14.3. Differential thermal analysis and thermogravimetry curves for polyethylene in nitrogen and oxygen. (From Kotoyori, Ref. 28.)

where the DTA and TG curves for the decomposition of polyethylene in air and nitrogen are shown. It is not until c. 400 °C that the weight loss begins in nitrogen, whereas in oxygen an exothermic reaction starts at c. 180 °C and the weight loss commences, corresponding to the first exothermic peak, at c. 230 °C. These facts suggest that in the presence of oxygen surface oxidation is occurring. Kinetic parameters for the degradation processes were calculated according to the method of Kissinger[29] (see Chapter 23).

Rudloff et al.[30] have investigated the thermal stability of a wide range of commercial polymers used in the construction of spacecraft and aeroplanes. Differential thermal analysis and thermogravimetry were applied to the reactions under atmospheres of 100% oxygen at 14.7 and 5 p.s.i. (101 and 34.5 KN m^{-2}) and in a mixture of 50% N_2 50% O_2 at 5 p.s.i. (34.5 KN m^{-2}). The volatile products were collected and then separated and identified by combined gas chromatography–mass spectrometry. Delrin® was shown to be relatively unstable and was degraded, commencing at *c.* 250°C in either atmosphere. Its major decomposition product is formaldehyde even under 1 atmosphere of oxygen. The primary degradation mechanism is apparently an 'unzipping' of the polymer and is unaffected by the presence of oxygen. Degradation of Mylar® is much more affected by availability of oxygen. From the data obtained it was concluded that attack begins on the ethylenic bridges. As the temperature is increased the residual ring compounds are involved in reaction, as is evidenced by the formation of acetylene and benzene type fragments.

APPLICATIONS TO THE STUDY OF POLYMERS FOR USE IN THE ELECTRONICS INDUSTRY

Polymers find a wide variety of uses in electronic parts and include printed circuit board materials, encapsulation or potting compounds, conformal coatings, wire insulations, etc. Some examples of the uses of DTA for studies of these materials are illustrated below (see Figs. 14.4 and 14.5).

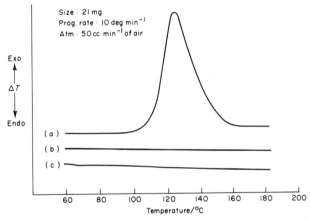

Fig. 14.4. Cure exotherm for Alkyd resin used in the manufacture of moulded connectors. (From Blaine, Ref. 31.)

Figure 14.4 illustrates how DTA (in this case DSC) may be used to check the degree of cure of an alkyd resin used in the production of moulded parts.[31] Curve (a) is for the uncured resin and shows the characteristic curing exotherm. Curve (b) is a re-run of the same sample after it had been cooled to the initial starting temperature. As can be seen from the lack of any peak, the resin is fully cured. If the material had been only partially cured, then this would have been shown by an exotherm somewhere between these two extremes. The bottom trace, curve (c), is the DTA trace obtained on a sample of a connector made from the same alkyd resin and shows that the resin is completely cured.

The preparation of printed circuit boards involves many steps. Initially the reinforcing material (glass, carbon, or other fibre) is impregnated with a thermo-setting resin such as an epoxy resin. This combination is then cured partially until it is no longer 'tacky'. The material is now in a form such that it can be handled easily and cut to the required shape and is referred to as 'prepreg' material. The final laminated board is manufactured by heating and pressing together several layers of prepreg. This last stage requires careful control of processing parameters to ensure that the board is fully cured. One method for determining the degree of cure of prepregs has been proposed by Fava[32] and involves measuring the glass transition temperature of the prepreg as a function of cure time. Figure 14.2 shows a DTA trace for two typical PCB materials, one with base resin and the other with a flame retardant added. Figure 14.5 illustrates the variation of T_g with

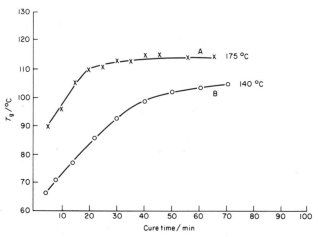

Fig. 14.5. Variation of glass transition temperature with cure time for the prepreg materials. (From Fava, Ref. 32.)

cure time for two other PCB materials, and clearly shows that prepreg A is completely cured after heating for 20 min at 175 °C, whereas prepreg B is not totally cured after heating at 140 °C for 70 min. This type of information is very useful for quality control of prepreg materials. Other examples of this type of study can be found in the work of Mantz and Creedon[33] and Creedon.[34]

Quality control of wire insulations and their resistance to oxidation is especially important when flammability properties are considered, and may be readily performed using DTA. The procedure used [35] is to maintain the sample in the DTA cell in an inert atmosphere at the required temperature until equilibrium is attained, as evidenced by a linear base-line. Reaction is initiated by switching the purge gas from the inert atmosphere to oxygen. Oxidation of the sample is observed as an exothermic deviation of the ΔT trace and the time taken for this to occur is taken as a measure of oxidative stability.

REFERENCES

1. H. W. Holden, *Proceedings of the First Toronto Symposium on Thermal Analysis* (H. G. McAdie, Ed.), Chemical Institute of Canada, Toronto, 1965, p. 113.
2. D. J. David, *Techniques and Methods of Polymer Evaluation* (P. E. Slade and L. T. Jenkins, Eds.), Marcel Dekker, New York, 1966, p. 43.
3. J. Mitchell and J. Chiu, *Anal. Chem.* **41**, 248R (1969).
4. D. A. Smith, *Differential Thermal Analysis*, Vol. 2 (R. C. Mackenzie, Ed.), Academic Press, New York, 1972, p. 379.
5. J. Chiu, *Polym. Prepr. Amer. Chem. Soc., Div. Polym. Chem.* **14**, 486 (1973).
6. B. Wunderlich, *J. Thermal Anal.* **5**, 117 (1973).
7. M. L. Dannis, *J. Appl. Polym. Sci.* **7**, 231 (1963).
8. R. H. White and T. F. Rust, *J. Appl. Polym. Sci.* **9**, 777 (1965).
9. T. S. Light, L. F. Fitzpatrick and J. P. Phaneuf, *Anal. Chem.* **37**, 79 (1965).
10. J. J. Keavney and E. C. Eberlin, *J. Appl. Polym. Sci.* **3**, 47 (1960).
11. R. F. Schwenker and R. L. Zuccarello, *J. Polym. Sci.* **C6**, 1 (1964).
12. B. Wunderlich, *J. Phys. Chem.* **64**, 1052 (1960).
13. R. A. Wersling, *Polym. Prepr. Amer. Chem. Soc., Div. Polym. Chem.* **14**, 283 (1973).
14. C. M. F. Vrouenracts, *Polym. Prepr. Amer. Chem. Soc., Div. Polym. Chem.* **13**, 529 (1972).
15. A. J. Kovacs, *Fortschr. Hochpolym. Forsch.* **3**, 394 (1963).
16. R. Miller and G. Sommer, *J. Sci. Instr.* **43**, 293 (1966).
17. B. Ke, *Newer Methods of Polymer Characterization*, Interscience, New York, 1964.
18. T. R. Manley and M. M. Qayyum, *Polymer* **13**, 587 (1972).
19. B. Wunderlich, *Thermochim. Acta* **4**, 176 (1972).
20. Du Pont Thermal Analysis Bulletin No. 900A-1.
21. B. Teitelbaum, *Dokl. Akad, Nauk SSSR* **169**, 1375 (1966).
22. M. A. Hughes and R. P. Sheldon, *J. Appl. Polym. Sci.* **8**, 1541 (1964).
23. N. D. Scott, *Polymer* **1**, 114 (1960).
24. B. Ke, *J. Polymer Sci.* **62**, 15 (1960).
25. H. C. Anderson, *Polymer* **7**, 193 (1966).
26. H. G. Langer and T. P. Brady, *Thermochim. Acta* **5**, 391 (1973).
27. R. J. Gaymans, K. A. Hodd and W. A. Holmes–Walker, *Polymer* **12**, 602 (1971).
28. T. Kotoyori, *Thermochim. Acta* **5**, 51 (1972).
29. H. E. Kissinger, *Anal. Chem.* **29**, 1702 (1957).
30. W. K. Rudloff, A. D. O'Donnel, R. G. Scholz and A. Valaitis, *Thermal Analysis* (*Proceedings of the Third Conference of Thermal Analysis*), Vol. 3, Birkhauser Verlag, Basle, 1972, p. 205.
31. R. L. Blaine, *Thermal Analysis in the Electronics Industry*, Paper presented at Educational Seminar Palo Alto CA June, (1974). Reprints may be obtained from Du Pont Co.
32. R. A. Fava, *Polymer* **9**, 137 (1968).
33. W. Manz and J. P. Creedon, *Thermal Analysis* (*Proceedings of the Third Conference of Thermal Analysis*), Vol. 3, Birkhauser Verlag, Basel, 1972, p. 145.
34. J. P. Creedon, Du Pont Thermal Analysis Brief 33 (1970).
35. J. B. Howard, *Polym. Eng. Sci.* **13**, 429 (1973).

CHAPTER 15

Coal and related Carbonaceous Materials

The first reported application of DTA to the study of coal appears to be that of Hollings and Cobb,[1] as early as 1914. Although the apparatus available at that time was relatively crude and the problems associated with the DTA of coal were, of course, not then appreciated, these workers nevertheless established the basic characteristic shape of coal curves. Subsequent papers lent support to the belief that the rank of a coal could be deduced from the shape of its DTA curve. Then in 1938, Kreulen et al.[2] appeared to have established a relationship between the ignition temperature of a coal and its volatile matter content, and hence with coal rank. Samples of powdered coal, volume about 1 cm³, were heated in flowing oxygen atmosphere and the temperature of onset of the exothermic reaction recorded; this ranged between about 200 °C and 260 °C, depending qualitatively on coal rank.

The work described above is now only of historical interest, due to lack of control of the experimental parameters. A much more rigorous and scientific approach to the DTA of coals was established by the work of Glass,[3,4] whose papers remain a significant contribution to coal science. Glass used 5 g samples of coal (100 mesh), heated at 10 °C min⁻¹ in static air, but contained in closed nickel vessels so that evolved gases would displace air from the vicinity of the samples. Oxygen was therefore excluded from the reaction zone, giving rise to conditions analogous to those in a coal carbonization plant. Coal samples used covered the entire range of types in normal production, but no attempt was made to isolate the petrographic constituents. The DTA curves obtained by Glass fell into five basic types, corresponding to the four natural boundaries in coal rank classification.

During the same period, King and Whitehead,[5] studied the DTA curves of certain coals under vacuum, but after separation of vitrinite from the remaining macerals. They were able to show that the major DTA peak given by the vitrinites increased from about 420–530 °C with increasing coal rank. Also, a pyridine extract of the vitrinites gave a similar peak to the vitrinite itself, while the solid residue seemed to be relatively unreactive.

However, subsequent studies by Berkowitz[6] and others[7] have not been able to reproduce the characteristic variations in the shape of the DTA curves as a function of coal rank. Many workers have therefore concluded that DTA is of little practical use as a method of assessing rank. While this may be true, nearly all the earlier work was carried out using DTA apparatus with the differential thermocouple junctions in contact with the decomposing coal. Furthermore, in much published data little care has been exercised in controlling and recording either the petrographic constitution of the coal samples or the nature and quantity of inorganic materials present. Additional measurements using modern DTA apparatus, with careful control of all the experimental variables, thus appear to be desirable. Differential thermal analysis data obtained for a range of vitrains by the present authors will be discussed later in this section.

The fields in which DTA can provide useful information on the structure, properties and behaviour of coals can be summarized as follows:

1. Coal rank classification.
2. Carbonization under various atmospheric conditions.
3. Determination of ignition temperatures.
4. Fluidized bed reactor studies.
5. Determination of the mineral content of coals.
6. Determination of carbonaceous materials present in other inorganic rocks.

Work on the DTA of coals published prior to 1955 has been reviewed by Nelson,[8] while more recent papers are discussed by Lawson,[9] in the book edited by Mackenzie, *Differential Thermal Analysis*. A comprehensive coverage of the whole field of coal science, which includes a section on DTA, is contained in the book[10] by Van Krevelen and Schuyer.

Reasons for irreproducibility

The reasons why DTA curves for coals, of an apparently similar chemical composition show such marked differences in practice, particularly when the results of different workers are compared, are not difficult to discern. Apart from differences in particle size and shape, packing density and other factors generally affecting DTA curves, with coals there are two additional and important variables. Firstly, coal samples frequently contain traces of minerals intimately mixed with the coal in such a way as to make complete separation virtually impossible. The chemical composition of these minerals will naturally vary from one location to another. Not only will these minerals themselves give rise to thermal effects on the DTA curve; they will also influence the mechanism of carbonization of the coal itself. This has been the subject of a number of investigations; see for example Refs. 11, 12 and 13.

The second important factor in that 'pure' coal itself consists of a mixture of petrographic constituents, the proportions of which vary. Each of these constituents from a given coal gives rise to DTA curves of the same general shape, but showing significant differences in detail.[14] Hence it is quite possible for the DTA curves given by the various macerals present in a certain coal to show differences as great, or greater than the differences observed between the curves from coals which are not of the same rank.

Reasons for the difficulty in interpretation of the DTA curves for coal during carbonization

This arises partly from the complex nature of the mechanism of the carbonization process and partly from the marked changes which occur in the physical properties of the solid residue after various stages of decomposition.

From the evidence available, it appears[15] that the mechanism of coal carbonization passes through the following stages, as the temperature is raised in an inert atmosphere. Up to about 150 °C, loss of physically adsorbed water predominates, causing little if any change in the coal structure. Then above 150 °C and below about 350 °C, a small amount of organic material, mainly alkyl aromatics, is evolved; this presumably arises from molecules trapped within the coal structure during the coalification process.

Primary carbonization commences at between 350 °C and 500 °C and involves fission of the carbon–carbon bonds in the aliphatic bridges which join together the aromatic clusters in the coal. As a result, some of the aromatic material present is liberated to form tar; hydrogen from the broken aliphatic bonds disproportionates between the residue and the tar, the latter being richer in hydrogen. Above 500 °C, the bulk of the non-aromatic material remaining in the residue appears to be hydrogen and methyl groups at the periphery of aromatic clusters.

Secondary carbonization involves the loss of this peripheral material, mainly in the form of methane and hydrogen, thus leaving the aromatic groups in the residue free to grow into sheets similar to, but much smaller than those in graphite. The mean layer diameter of these sheets appears to reach a steady value by about 800 °C; little further growth occurs until temperatures well above 1000 °C.

The greatest obstacle to the interpretation of the DTA curves is their pronounced and variable degree of base-line drift. Since this is due to the nature of the decomposition process itself, it cannot be corrected by electrical feedback or by careful matching of the sample and reference material. Throughout most of the carbonization process, decomposition products are being evaporated from the solid residue, although the amount and chemical constitution of the evolved material varies enormously over the temperature range. Hence the DTA curve is continually being influenced by the enthalpy of vaporization of the evolved decomposition products.

A second cause of base-line drift seems to be the marked changes in thermal conductivity[16] and other physical properties such as porosity and plasticity, which occur in the residual material during carbonization.

CARBONIZATION AND COAL RANK CLASSIFICATION

The problems associated with using DTA for the determination of coal rank, or for studying the mechanism of the carbonization process, will be apparent from the two preceding sections. This is not to say that DTA cannot give valuable information, but rather that it should only be used in conjunction with other techniques. Control of the experimental conditions, and of the sample in particular, is more than usually important.

For optimum DTA curves, the sample should contain one petrographic constituent only; vitrinite appears to give the most characteristic and reproducible results. The mineral content must, of course, be kept to a minimum. It is suggested that the sample be ground to −72 BS mesh and that coke or char originating from a similar source be used as the reference material. An inert atmosphere is essential, for obvious reasons. The heating rate does not appear to be important, so that the usual conditions of 10 °C min⁻¹ are generally found to be convenient. Sample dilution has been employed by several workers in order to reduce base-line drift, but it involves the risk of influencing the carbonization process, particularly where inorganic diluents are used.

Differential thermal analysis curves have been determined by the authors for vitrain samples representative of the entire coal range; details of these are given in Table 15.1. Measurements were carried out using −72 BS mesh samples

TABLE 15.1

Details of the vitrain samples used by the authors in their DTA studies.

Origin of Vitrain	% Carbon (d.a.f.) (BS 1016/6)	% Volatile matter (BS 1016/3)	National Coal Board Rank No.
Nailstone, Yard seam	78.5	43	902
Cannock Wood, Shallow seam	80.0	43	802
Markham Main, Barnsley seam	82.5	41	602
Dinnington Main, Barnsley seam	85.0	37	502
Houghton Main, Parkgate seam	87.5	32	401
Roddymoor, Ballarat seam	88.5	27	301
Coegnant, Gellideg seam	91.5	16	203
Blaenhirwaun, Pumpquart seam	94.0	5	101
Castlecomber, Skehena seam	96.0	2	—
Natural graphite	> 99	< 1	—

weighing approximately 90 mg, with flowing nitrogen atmosphere and a heating rate of 10 °C min⁻¹. Coke breeze of similar mass and particle size was chosen as the reference material.

Because of the relative complexity of the curves obtained, only the major features will be discussed here. In all cases there were three main peaks, the first occurring between room temperature and about 200 °C. This peak is associated with loss of adsorbed water and other low molecular weight volatile material. As might be expected, this peak was found to be endothermic with all the samples, with the peak area falling progressively as the coal rank increased. The reduction in peak area thus reflects the fall in specific surface and porosity of the vitrains[17] with increasing coal rank.

The second major peak, corresponding to the primary carbonization process, shows the greatest variation with coal rank. For lower rank coals, having carbon contents below about 90%, this peak is always exothermic and the peak onset temperature increases with rank (Fig. 15.1). Such a finding is in accordance with

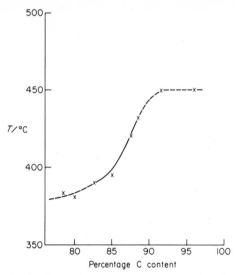

Fig. 15.1. Primary carbonization peak onset temperature as a function of carbon content.

the work of Fitzgerald and Van Krevelen,[18] who have shown that the energy of activation for gas evolution during primary carbonization increases progressively from 9.5 J g^{-1} at 80% carbon content, to 14.3 J g^{-1} for a carbon content of 92%.

For vitrains having carbon contents between about 90% and 94%, primary carbonization gives rise to a combination of exothermic and endothermic effects on the DTA curves, which show various inflexions. Under these circumstances it is difficult to distinguish any clearly defined, separate peaks. Then with meta anthracite and natural graphite, having carbon contents of 96% or above, the primary carbonization peak is undoubtedly endothermic, although it is very broad and indistinct in the case of graphite.

The primary carbonization peak thus reflects the result of an exothermic decomposition reaction accompanied by endothermic volatilization of the tar which has been formed. Now the mass of tar evolved can be determined from TG data obtained under comparable conditions. If the enthalpy of volatilization of the tar is taken[19] as 336 J g^{-1}, assumed to be independent of rank, then the endothermic contribution to the peak area can be calculated. If this correction is made to the primary carbonization peak area, it is found that the enthalpy of decomposition per milligram of tar formed is more or less constant for coals of 90% carbon content and below. Such a result must imply that the mechanism of the primary carbonization process is similar for all coals below the rank of anthracite.

Secondary carbonization gave rise to an exothermic peak in all cases, with the peak onset temperature remaining approximately constant for vitrains below the rank of anthracite (Fig. 15.2). Thereafter the onset temperature increased progressively with increasing carbon content. This no doubt reflects the growth in mean layer diameter of the aromatic sheets in anthracites, since Hirsch[20] has shown that coals below about 90% carbon content contain aromatic layers

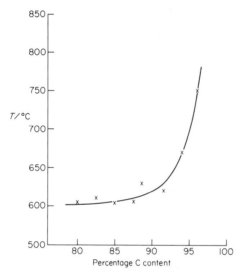

Fig. 15.2. Secondary carbonization peak onset temperature as a function of carbon content.

corresponding to an average size of between two and four condensed benzene rings. At higher carbon contents, the calculated mean layer diameter increased rapidly, thereby leading to increasing thermal stability of the peripheral groups which must be lost during secondary carbonization.

Clearly a detailed interpretation of DTA curves obtained with coals must lie outside the scope of a book such as this, since the subject is one of considerable complexity. Nevertheless, it is hoped that the foregoing discussion will have served as a guide to the types of curve which may be expected.

DETERMINATION OF IGNITION TEMPERATURES

Basically, this involves heating a powdered sample of the coal, or carbonaceous material, in a flowing oxygen (or air) atmosphere and recording the onset temperature of the resulting exotherm. The heating rate does not appear to be important and an empty bucket can be used in the reference cell. Merrill[21] has used this method with a static atmosphere, at pressures up to 420 lbs in^{-2}. Under these conditions there was sometimes a lag of up to 100 °C between the first observable deviation of the DTA trace from the base line and occurrence of ignition. Peak onset temperatures were found to range between 160–300 °C, while ignition temperatures ranged from 180–400 °C.

Dollimore et al.[22] studied the effect of inorganic additives on the ignition temperature of carbons in flowing air. Small samples (c. 5 mg) were found to minimize base-line drift and the peaks were sharp enough for the extrapolated onset temperature to give a measure of ignition temperature; values of between 703 °C and 868 °C were obtained.

FLUIDIZED BED STUDIES

In order to produce fluidization, a stream of gas has to flow through the sample at a rapid and controlled rate. This requires a DTA cell of the type used in the study of solid catalysts, such as is described elsewhere (see Fig. 11.1) in this book. Relatively little work has been published in this field, but the way in which DTA can be employed for such studies is obvious. It is suggested that the reader refer to the Chapters dealing with solid catalysts and with gas–solid reactions (8 and 11).

DETERMINATION OF THE MINERAL CONTENT OF COALS

The amounts of various clay and other minerals present in coals have been estimated by the height, or area under their characteristic DTA peaks. This method is not very accurate and even the identity of the minerals themselves may be suspect. Worrall[23] has pointed out that the peak, which normally occurs at 600 °C in kaolin, appears to have shifted to 580 °C for kaolin incorporated in coal. Accordingly, the method ought not to be used without a serious consideration of all possible sources of error (see Chapter 9).

DETERMINATION OF CARBONACEOUS MATERIALS PRESENT IN OTHER MINERALS

This method is applicable where the host mineral is thermally stable at moderate temperatures. Combustion of the carbonaceous material in a stream of flowing oxygen then gives rise to an exotherm on the DTA curve, the area under the peak being a measure of the amount present. Clearly the heat of combustion of the carbonaceous material must be known for this calculation to be made. The principle of the method is identical with that described for the determination of adsorbed films of organic substances on inorganic substrates (Chapter 10), elsewhere in this book.

WOOD

Logically the first stage of the coalification process, which leads ultimately to graphite, is natural wood. Interest in the thermal decomposition of wood has been concerned mainly with imparting fire resistance, through various forms of chemical treatment. An excellent account of the use of DTA in such studies, illustrated with many typical DTA curves, has been given by Tang.[24]

The mechanism of carbonization of wood in an inert atmosphere follows a very similar course to that of low rank coals, although more volatile matter is lost.[28] It has been established that the DTA curves given by wood represent the combined effects due to the main components of the wood, i.e. lignin, cellulose and xylan. This is well illustrated by the work of Shafizadeh and McGinnis[25] shown in Fig. 15.3; the curves are for cottonwood and its components, heated at 15 °C min^{-1} in nitrogen, using 5 mg samples and silica beads as the reference material.

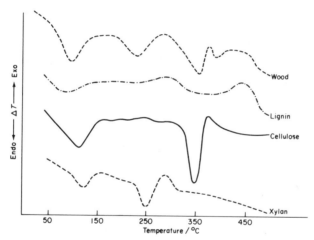

Fig. 15.3. Differential thermal analysis of cottonwood and its components (From Shafizadeh and McGinnis, Ref. 25.)

PAPER

Differential thermal analysis has been used in an attempt to characterize papers and modified cellulose materials.[26, 27] Two main endothermic peaks were observed. The first, occurring between about 75 °C and 150 °C, has been ascribed to loss of moisture and other adsorbed substances. The main decomposition endotherm occurs, in an inert atmosphere, between about 300 °C and 380 °C and is associated with formation of a tar consisting mainly of levoglucosan. Some typical DTA curves for paper are illustrated in Fig. 15.4.

The procedure for sample preparation used by several workers is as follows. About 25 mg paper is weighed into the DTA sample container and a few drops

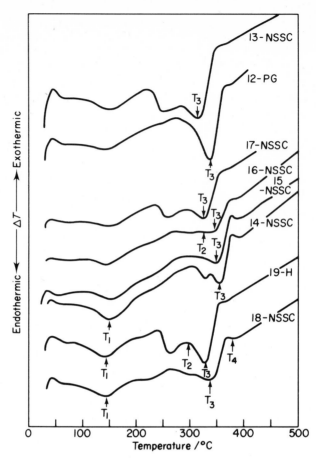

Fig. 15.4. Differential thermal analysis traces of some commercial printing papers (From Herbert, Tryon and Wilson, Ref. 26.)

of water added; the paper is then macerated to a pulpy consistency. The container should be placed in an oven at about 100 °C and air-dried to constant weight, before being cooled and placed in the DTA apparatus. A heating rate of 20 °C min⁻¹ has generally been employed, in conjunction with a flowing nitrogen atmosphere. The reference material can either be alumina or powdered glass, previously heat-treated to remove any volatile matter.

REFERENCES

1.   H. Hollings and J. W. Cobb, *J. Gas Light.* **126**, 917 (1914).
2.   D. J. W. Kreulen, C. Krijgsman and D. T. J. Horst, *Fuel (London)* **8**, 243 (1938).
3.   H. D. Glass, *Econ. Geol.* **49**, 294 (1954).
4.   H. D. Glass, *Fuel (London),* **34**, 253 (1955).

5. L. H. King and W. L. Whitehead, *Econ. Geol.* **50**, 22 (1955).
6. N. Berkowitz, *Fuel (London)* **36**, 355 (1957).
7. M. F. Kessler and H. Romovackova, *Fuel (London)* **40**, 161 (1961).
8. J. B. Nelson, British Coal Utilization Research Association. Review 152 (1955).
9. G. J. Lawson in *Differential Thermal Analysis*, Vol. 1 (R. C. Mackenzie, Ed.), Academic Press, London, 1970, Chapter 25.
10. D. W. Van Krevelen and J. Schuyer, *Coal Science*, Elsevier, Amsterdam, 1957.
11. A. F. Gaines and R. G. Partington, *Fuel (London)* **39**, 193 (1963).
12. F. Yoshimura, S. Mitsui and T. Mitani, *J. Fuel Soc. Jpn* **44**, 575 (1965).
13. F. Yoshimura and S. Mitsui, *J. Fuel Soc. Jpn* **45**, 191 (1966).
14. C. Kroger and A. Pohl, *Brenstoff. Chem.* **38**, 179 (1957).
15. M. I. Pope, *Polymer* **8**, 49 (1967).
16. A. F. Boyer and P. Payen, Third International Conference on Coal Science, Valkenburg, (1959).
17. S. J. Gregg and M. I. Pope, *Fuel (London)* **38**, 501 (1959).
18. D. Fitzgerald and D. W. Van Krevelen, *Fuel (London)* **38**, 17 (1959).
19. Coal Tar Research Association, Research Report 0131 (1955).
20. P. B. Hirsch, *Proceedings of the Conference on Science in the Use of Coal, Sheffield*, Institute of Fuel, London, 1958, pp. A29–A33.
21. V. S. Merrill, *Fuel (London)* **52**, 61 (1973).
22. D. Dollimore, L. F. Jones and T. Nicklin, *Thermochim. Acta* **5**, 265 (1973).
23. W. E. Worrall, *J. Inst. Fuel* **38**, 280 (1965).
24. W. K. Tang, in *Differential Thermal Analysis*, Vol. 2 (R. C. Mackenzie, Ed.), Academic Press, London, 1972, Chapter 45.
25. F. Shafizadeh and G. D. McGinnis, *Carbohydr. Res.* **16**, 273 (1971).
26. R. L. Herbert, M. Tryon and W. K. Wilson, *J. Tech. Assoc. Pulp and Paper Ind.* **52**, 1183 (1969).
27. E. J. Parks, *J. Tech. Assoc. Pulp and Paper Ind.* **54**, 537 (1971).
28. F. Shafizadeh, K. V. Sarkanen and D. A. Tillman, *Thermal Uses and Properties of Carbohydrates and Lignins*, Academic Press, New York, 1976.

CHAPTER 16
Petroleum and Related Products

A considerable amount of work has been carried out in recent years on the use of DTA as a means of characterizing petroleum products. One major aim of this research has been to develop simple tests which give data of fundamental significance, as an alternative to the empirical procedures laid down in existing test specifications. Many of the latter involve the use of complex, purpose-built apparatus, are slow to carry out and require relatively large test samples. Furthermore, the results given by many of these standard procedures are of comparative value only, since the data cannot easily be correlated with the physical and chemical properties of the test substance.

Much of the progress made in establishing the suitability of DTA as an alternative test procedure is due to the work of Noel[1, 2] in Canada and of Giavarini and Pochetti[3] in Italy, although both actually used DSC for their experimental measurements. Noel has listed the advantages of using DTA/DSC as follows:

1. The measurements are quick to perform and reproducibility is high.
2. Only very small samples, < 10 mg, are normally required.
3. The equipment is suitable for many laboratory purposes, while the conventional test procedures require a specific piece of apparatus for each test.
4. The data obtained are of fundamental physical, or chemical significance.

It will, of course, be necessary to employ a DTA apparatus, which is capable of controlled cooling, as well as controlled heating, since many of the transitions involved will occur at subambient temperatures. For several tests, it has been found advantageous to work under positive pressures of several,[1, 4] or even tens[5, 6] of atmospheres. In any case, lubricating oils and particularly hydraulic fluids, frequently work under high hydrostatic pressures; therefore it will be necessary to simulate those conditions in the DTA cell for the tests to have any real value. Careful choice of a suitable apparatus is thus essential for a laboratory working with automotive hydrocarbon products.

Fields in which DTA has proved to be applicable to the study of petroleum and related products include the following:

1. Characterization and thermal stability of lubricating oils.
2. The efficiency of oxidation inhibitors added to lubricating oils, fuel oils and light hydrocarbon fuels.

3. The temperature–pressure stability of hydraulic fluids.
4. The identification of waxes, and their transition temperatures.
5. The characterization and wax content of bitumens.

LUBRICATING OILS

Characterization tests

Lubricating oils may contain linear and branched hydrocarbons; on cooling, the former readily crystallize to form a wax, while the latter either do not crystallize at all, or do so only very slowly. Hence a DTA curve might be expected to show a glass transition temperature, associated with the non-crystalline components in the oil, and enthalpy changes resulting from the latent heat of fusion of the wax content.

Now the wax crystallization temperature of an oil is one of the characteristic properties determined in standard test specifications and is referred to as the *cloud point*. This temperature can equally well be obtained from the onset of the melting endotherm observed during DTA. In addition, the temperature at which the curve returns to the base-line will give the range of temperature over which the wax melts. Since the standard test specification involves observation of the oil during slow, controlled cooling, DTA gives a far more objective record of the cloud point and is also much quicker. A cooling–heating rate of 10 °C min⁻¹ appears to be perfectly suitable for this purpose. Some typical DSC curves obtained by Noel are illustrated in Fig. 16.1. The correlation between cloud point measurements made according to the ASTM test specification and those taken

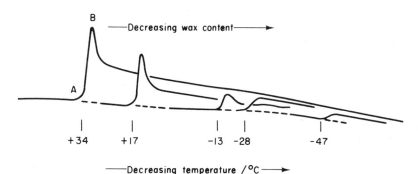

Fig. 16.1. Some cooling curves for lubricating oil showing the effect of wax content; cooling rate 10 °C min⁻¹. (After Noel, Ref. 2.)

from the onset temperature (A) of the DSC cooling exotherm is demonstrated in Fig. 16.2.

A second important characteristic used to define a lubricating oil is its *pour point*. This can be regarded as the temperature below which wax crystals are present in a sufficient amount to form a gel-like structure. Free flow of the oil is thereby inhibited and may lead to blocking of filters. The pour point of oils having a similar molecular structure would therefore be expected to depend on their wax content. Noel has proposed that the exothermic peak temperature

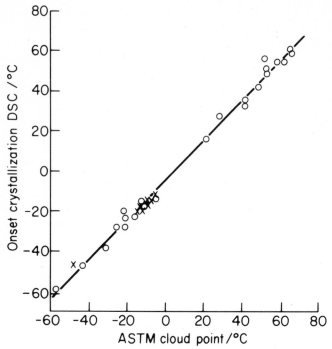

Fig. 16.2. Correlation of the ASTM cloud point and the DSC crystallization onset, using DSC cooling (10 °C min⁻¹) and the ASTM cloud point test. (x), fuel oils (2 types); (O), lube oils (11 types). (After Noel, Ref. 2.)

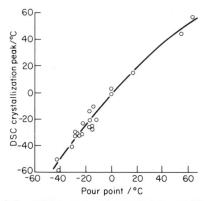

Fig. 16.3. Correlation of the ASTM pour point and the crystallization peak temperature, using DSC cooling (10 °C min⁻¹) on six different lube oils. (After Noel, Ref. 2.)

(B, Fig. 16.1) due to wax crystallization may give a measure of the point at which sufficient wax has solidified to cause oil caging. If this is so, then it should be possible to estimate the pour point from a DTA/DSC cooling curve. The correlation found by Noel[2] is shown in Fig. 16.3.

As mentioned previously, a lubricating oil will normally also contain constituents that do not form waxes on cooling. This material will become progressively more viscous as the temperature falls, until it eventually undergoes a *glass transition*; negligible heat is evolved or absorbed during this transition, but the material undergoes a marked change in heat capacity. Hence the glass transition shows up as a point of inflexion, or sharp change in slope of the corresponding DTA/DSC curve. Attempts have been made[7] to predict the temperature viscosity index of lubricating oils from their glass transition temperatures.

Thermal degradation

The useful working life of a lubricating oil is determined primarily by its thermal stability in the presence of air. It has been suggested[1, 3] that the ASTM oxidation test, which measures resistance to degradation, might well be replaced by DTA/DSC measurements.

On heating a sample of oil in air, using a DTA apparatus, the base line initially tends to show a slight endothermic drift due to evaporation of more volatile material. This is followed by an exothermic effect resulting from oxidative degradation of the oil; the onset temperature of the exotherm can be taken as a measure of thermal stability. Some typical DSC curves are illustrated in Fig. 16.4.

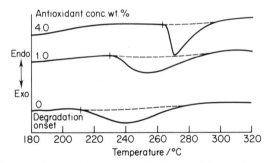

Fig. 16.4. Effect of antioxidant on lube oil stability. Differential scanning calorimetry analysis 100 psig. (After Noel, Ref. 1.)

Krawetz and Tovrog found[4] that the onset of the exothermic reaction became sharper when the measurements were made at an elevated pressure. Since lubricating oils usually operate under pressure anyway, this seems to be an added justification for pressurising the DTA apparatus.

Various workers have used this technique to study the effect of a range of oxidation inhibitors on lubricating oils. The higher the onset temperature of the exothermic reaction, the more effective the inhibitor, and conversely. For example,

Hopkins[8] investigated a number of metal-organic compounds, including naph-thenates, octoates, carbonyls, etc., using a Stone differential thermal analyser and a flowing oxygen atmosphere, with a heating rate of 11 °C min^{-1}. The oil samples containing the dissolved inhibitor weighed approximately 66 mg. He established that the effect of the added metal was almost independent of the structure of the compound containing it. Whereas oxidation was strongly inhibited by copper compounds, it was actually promoted by cobalt-containing additives.

An alternative approach[12] is to heat the test sample at a constant temperature, in a flowing oxygen atmosphere, and record the induction time before oxidation commences.

Clearly DTA can prove very useful as a screening test to evaluate the efficiency of various oil additives in inhibiting oxidation, and in determining the concentrations required.

FUEL OILS

What has been said about lubricating oils applies in almost all respects to fuel oils.

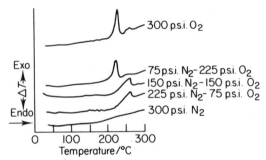

Fig. 16.5. Effect of varying oxygen content at constant pressure on oxidation peak of ASTM iso-octane. (After May and Bsharah, Ref. 6.)

LOW MOLECULAR WEIGHT HYDROCARBONS: PETROL

Here the main application of DTA is to study the efficiency of oxidation inhibitors. Measurements can be made[6] in a closed system under a high pressure of oxygen and the temperature at which the exothermic reaction is first detectable recorded. Since oxidation sometimes starts off rather slowly, other significant points on the DTA curve are the extrapolated onset and the peak temperatures. A typical example is illustrated in Fig. 16.5. Once again, this technique may be used as a screening test of oxidation inhibitors.

HYDRAULIC FLUIDS

Resistance to thermal degradation under pressure is a particularly important requirement for hydraulic fluids, especially those intended for use in braking systems. Levy *et al.*[5] have demonstrated the use of DTA/DSC for assessing the quality of brake fluids. Differences between three samples did not show up very well from DSC curves obtained in air, or nitrogen, at ambient pressures. By raising the pressure to 600 lb in⁻², the onset temperatures of the degradation reactions became much clearer, thereby facilitating selection of the most stable fluid. The relevant curves are shown in Fig. 16.6.

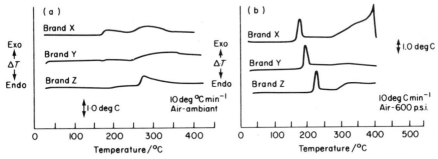

Fig. 16.6. Differential scanning calorimetry curves automotive brake fluid, brands X, Y and Z, (a) at atmospheric pressure (b) at 600 lb in⁻². (After Levy *et al.*, Ref. 5.)

WAXES

Waxes can be obtained from the distillation of crude mineral oil, by the extraction of vegetable and animal products, or by direct synthesis of organic polymeric materials. Whatever the source, waxes consist of a complex mixture of high molecular weight organic compounds, which are extremely difficult to separate. No doubt this is partly the reason why waxes are often described by their appearance, or origins, rather than by their chemical constitution. Differential thermal analysis has proved to be of considerable value in identifying waxes and of estimating one type of wax in the presence of another. A comprehensive study, using DSC, of waxes from a wide variety of sources has been described by Flaherty;[9] both characteristic curves and thermodynamic data obtained from them are discussed.

Petroleum waxes

Waxy hydrocarbon constituents are present in many crude oils and, on distillation, appear either in the lubricating oil fraction or in the residue. In the former case, as much wax as possible has to be removed, for the reasons previously discussed. The wax contained in the lube distillate fraction shows a high degree of crystallinity and is known as paraffin wax, while that found in the residue is termed microcrystalline wax.

Paraffin waxes are characterized by showing at least two endothermic effects on heating to a temperature just above the melting point. The lower temperature peak has been ascribed to an orthorhombic-hexagonal solid state phase transition, while that occurring at the higher temperature is, of course, due to the latent heat of fusion. Currell and Robinson[10] found that with all the paraffin waxes they studied, the DTA curve had returned to the base-line by 460 °C.

Microcrystalline waxes, on the other hand, were found to give rise to only one low-temperature endotherm, corresponding to the melting point. On further heating to 600 °C, Currell and Robinson reported that all their microcrystalline waxes gave a characteristic endotherm with a peak temperature of about 475 °C to 480 °C. The area of this peak was used to determine the amount of microcrystalline wax present in a mixture with paraffin wax.

However, the interpretation of DTA curves of waxes should be treated with caution, since Flaherty gives[9] a curve for microcrystalline wax which clearly exhibits at least two low temperature endothermic effects on heating. After cooling and subsequent re-heating, only one broad endotherm was obtained. Since all the microcrystalline wax samples used by Flaherty behaved in this way, it appears to be necessary to pre-melt the samples before using their DTA/DSC curves to differentiate between paraffin and microcrystalline waxes.

The differences between the thermal behaviour of paraffin and microcrystalline waxes arises from differences in their molecular constitution. Paraffin waxes are composed[3] mainly of saturated, normal paraffins, containing between about 18 and 40 carbon atoms per molecule. Microcrystalline waxes consist mainly of branched chain paraffins of higher molecular weight, typically having between 25 and 65 carbon atoms per molecule.

Giavarini and Pochetti[3] have used the melting endotherm of paraffin waxes to determine the wax content of lubricating oil fractions. As the proportion of oil to wax increases, so the two low-temperature endothermic effects tend to merge. The total area of this combined endothermic peak can be used to determine the wax content of the oil, from the calibration curve shown in Fig. 16.7.

Polyethylene waxes appear to behave in a very similar way to microcrystalline waxes, in that the heating curve reveals some minor endothermic peaks, which are destroyed by cooling and subsequent re-heating.

Animal and vegetable waxes

Natural waxes are far more complex in their structure and compositions than are petroleum waxes. In addition to paraffins and alcohols, these waxes also contain varying amounts of carboxylic esters, free acids and hydroxy acids, ketones and in some cases, sterols. This complexity is, not unexpectedly, reflected by their DTA curves. Hence individual waxes can be identified by comparing their DTA curves with those given by samples of known composition and purity. Alternatively, the test sample can be run against a reference sample consisting of a wax believed to be of the same composition. If the assumption is correct, then no significant thermal effects should appear on the resulting DTA curve.

Nevertheless, it would be wise in all cases of doubt to confirm the identity of a

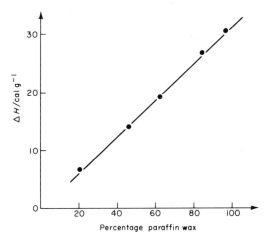

Fig. 16.7. Melting ΔH plotted against paraffin wax content of lubricating oil. (From Giavarini and Pochetti, Ref. 3.)

TABLE 16.1

Total enthalpy changes and transition temperatures of some natural and hydrocarbon waxes. (After Flaherty, Ref. 9.)

Wax	Total enthalpy change/cal g^{-1}		Transition temperature/°C	
	Mean	Range	Liquid–solid transition	Solid–solid transition(s)
Paraffin, m.p. 46–60 °C	32.1[a]	28.4–35.2	37–58	—
Paraffin, m.p. 62–75 °C	43.2[b]	41.0–45.5	60–65	50–53
Microcrystalline	32.3	28.8–38.6	70–100	—
BP slack wax	16.1	15.0–16.5	57–58	—
Polyethylene	12.4[c]	8.5–14.7	80–90	—
Montan	28.2	25.2–31.6	71–78	53–65
Ceresin	31.8	31.3–32.6	50–56	—
Peat wax	13.5	13.4–13.6	57–63	—
Ozocerite	31.0	30.5–31.4	72–73	60–62
Carnauba	38.7[b]	37.0–40.7	80–82	72–73
Candelilla	32.1	30.3–34.8	64–65	54–55
Beeswax	32.0[b]	31.1–32.8	60–63	54–55, 50–51
Esparto	30.2	29.8–31.3	66–67	50–51
Spermaceti	31.0	30.5–31.3	—	—
Sugar cane	35.4	33.6–37.2	73–75	60–63
Japan wax	21.5	20.8–22.2	34–35	—
Raffia wax	23.6	23.4–23.8	74–75	—
Ouricury	31.3	29.8–32.8	70–71	—
Shellac	38.5	36.4–39.7	70–71	60–61, 48–50, 35–37

[a] Mean of determination on 16 different samples.
[b] Mean of determination on 6 samples from different sources.
[c] Mean of determinations on 5 AC–polyethylene samples (Allied Chemical Corp.) having molecular weights in the range 2 000–7 000.

wax by infrared absorption spectroscopy, again by comparing spectra from test and reference materials. Details of the latent heat of fusion and melting temperature range of representative waxes are recorded in Table 16.1, taken from the work of Flaherty.

BITUMEN

Bitumens, otherwise known as asphalts, represent the least volatile products of crude oil distillation. They consist of a mixture of both aliphatic and aromatic hydrocarbons of very high molecular weight, together with a certain amount of waxy material. A method of determining the wax content, very similar to that suggested for lubricating oils, has been published by Giavarini et al.[11] Basically, this involves measuring the area under the endotherm due to the enthalpy of fusion of the bitumen. The value obtained is then converted to wax content using a calibration graph, such as that shown in Fig. 16.8.

There also appears to be a correlation between the glass transition temperature of bitumens and their brittle point, as measured by the Fraass test.

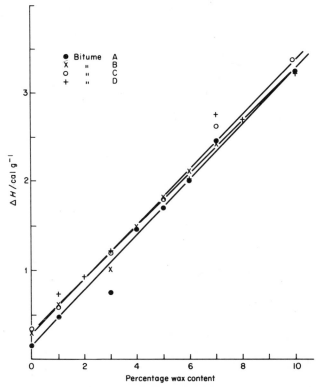

Fig. 16.8. Latent heat of fusion of bitumen A, B, C and D plotted against wax content. (From Giavarini et al., Ref. 11.)

REFERENCES

1. F. Noel, *J. Inst. Petrol* **57**, 354 (1971).
2. F. Noel, *Thermochim. Acta* **4**, 377 (1972).
3. C. Giavarini and F. Pochetti, *J. Thermal Anal.* **5**, 83 (1973).
4. A. A. Krawetz and T. Tovrog, *Ind. Eng. Chem., Prod. Res. Dev.* **5**, 191 (1966).
5. P. F. Levy, G. Nieuweboer and L. G. Semanski, *Thermochim. Acta* **1**, 429 (1970).
6. W. R. May and L. Bsharah, *Ind. Eng. Chem., Prod. Res. Dev.* **10**, 66 (1971).
7. R. S. Stearns, I. N. Duling and R. H. Johnson, *Ind. Eng. Chem., Prod. Res. Dev.* **5**, 306 (1966).
8. P. D. Hopkins, *Ind. Eng. Chem., Prod. Res. Dev.* **6**, 246 (1967).
9. B. Flaherty, *J. Appl. Chem. Biotechnol.* **21**, 144 (1971).
10. B. R. Currell and B. Robinson, *Talanta* **14**, 421 (1967).
11. C. Giavarini, F. Pochetti and C. Savu, *Riv. Combust.* **25**, 149 (1971).
12. G. E. Cranton, *Thermochim. Acta* **14**, 201 (1976).

CHAPTER 17

Fibres

In this chapter we are concerned with the use of DTA in the studies of fibrous materials. These include such materials as textiles, fibre glass and carbon fibres. Fibres are generally formed from natural and synthetic high polymers; hence this is essentially an extension of the chapter on polymers (Chapter 14). In essence, much of what was said there is equally applicable to fibres, so DTA has been used to study a similar range of processes. These include physical transformations on heating, thermal decomposition characteristics, effects of chemical and/or physical pretreatment and the influence of process conditions. For a detailed review on the applications of DTA to fibres the reader is directed to the excellent review by Schwenker and Chatterjee.[1]

EXPERIMENTAL TECHNIQUE

Small samples (c. 5–20 mg) are to be preferred, where it is known that the sample is truly homogeneous, in order to minimize thermal gradients within the sample. Small samples also facilitate the release of volatile degradation products; this may well reduce the effects of secondary reactions between the various gaseous products. Problems may well be encountered in preparation of the sample (see Chapter 14). Schwenker et al.[2] recommend that the fibres be cut into small pieces some two to three millimetres in length. If grinding is necessary, it should be carried out to the minimum extent possible, since the heat generated by the grinding process will probably affect the sample. This is especially important if low-temperature DTA is to be used.

It may well prove necessary to dilute the sample, in a similar manner to that used in conventional polymer studies. Both layered samples and mixtures containing sample concentrations of up to 30 % have been used. These approaches have the advantage of matching more closely the heat capacities of the sample and the reference material and should go a long way to reducing the effects of drifting base-lines. However, care must be exercised to ensure that there is no reaction between the sample and the diluent material. In cases where very small sample weights are used (< 5 mg) it may well prove advantageous to use an empty crucible as the reference cell. In general the standard heating rate of

$10\,^{\circ}\text{C min}^{-1}$ would seem to be the most widely used and offers the best compromise between run time and sensitivity.

TYPICAL DTA CURVES

Let us now consider a specific example that has been cited[2, 3] in the literature in order to illustrate the types of effect that can be determined by DTA. Figure 17.1 shows the DTA curve obtained on a sample of undrawn Dacron® (polyethylene terephthalate) fibre when heated under a nitrogen atmosphere. The similarity of this curve to Fig. 14.1 (p. 107) should be readily apparent. The effect which is shown occurs at a temperature of about 77 °C and is marked by a pronounced change in the slope and position of the base-line; this can be attributed to a glass transition. For a precise measurement of the temperature at which this change

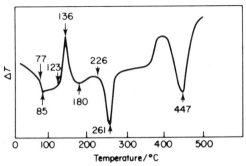

Fig. 17.1. Differential thermal analysis curve for undrawn Dacron® fibre in nitrogen. (From Beck and Schwenker, Ref. 3.)

occurs it would be necessary to use the method outlined in Chapter 5. Following this transition, a relatively sharp exotherm is obtained with a peak temperature of 136 °C and is due to polymer crystallization. The presence or otherwise of this peak is dependent on the history of the sample and is frequently used in this context.

The endotherm at 260 °C is due to melting of the polymer. The results described so far would be self-evident to someone familiar with the DTA of polymeric materials but would be of little value to the beginner. In the latter's case it would be necessary to run a TG experiment in order to show that these effects are accompanied by little or no loss in weight of the sample and must therefore be due to physical effects. Similarly an experiment with a hot stage microscope would show the effects of melting of the sample. This is mentioned here in order to stress the necessity of interpreting a DTA trace in conjunction with data from other experimental techniques. Above the melting temperature, a complex heat effect is observed with competing exothermic and endothermic reactions occurring. Since the atmosphere used in this investigation is inert, then these exothermic reactions are attributed to the degradation process itself and not to subsequent reactions with atmospheric oxygen. Let us now consider a few examples where DTA has been used to look at fibres.

Glass transitions

It has previously been mentioned that DTA offers almost the best method for the determination of glass transition temperatures. The very high sensitivities available in present day equipment enables the small heat effects associated with glass transitions to be detected relatively easily.

Haly and Snaith[4] have studied the glass transition of wool by DTA using relatively large samples of between 100 and 250 mg. The samples were placed in capsules and sealed in with varying amounts of water. The temperature at which the transition occurred was shown to be dependent on the amount of water present, the heating rate and the type of keratin fibre. For instance, for a vacuum-dried Merino wool, the transition occurred at around 210 °C whereas for the same wool in the presence of c. 7 % water the transition occurred at a temperature closer to 190 °C.

The effects of cooling molten samples of Dacron® fibres on the transition temperature have been investigated by Schwenker.[3] An example of the results obtained is that the transition temperature is reduced to 71 °C when the molten sample is quenched in a solid carbon dioxide–methanol mixture. This is in contrast to a temperature of 77 °C for the original fibre.

Crystallization and melting points

There is little to be added to the description given for polymers on pp. 107–109. Examples of the uses of DTA in this area of study can be found in the investigations into polyethylene terephthalate,[5] nylon[6] and cellulose.[7]

Decomposition

Differential thermal analysis has been widely used to study the thermal stability of fibres and particular importance is placed upon this type of evaluation in connection with the fire properties of textile materials. Cellulose[8, 9] has been shown to degrade exothermically, commencing at about 260 °C, with a peak maximum at around 320 °C. This has been attributed to the auto-oxidation of carbonyl groups and C—H bonds. Superimposed upon this process is a strong endothermic peak due to the heat absorbed by degradation and chain scission. Above about 370 °C a strong exotherm is obtained due to oxidative processes, such as carbonization to form condensed ring structures. Differential thermal analysis for the process under a nitrogen atmosphere is similar, except that this latter exotherm is absent, showing that it is due to reactions with atmospheric oxygen.

Orlon® presents a complex degradation pattern and serves as a good example of the uses of TG as a complementary technique to DTA. It has been known for many years that if a sample of Orlon® is heated up to 250 °C in air it undergoes a colour change from light yellow to black. This so-called 'Black Orlon' is insoluble in normal solvents and the fibre is resistant to further heating. The precursor to Orlon®, polyacrylonitrile, has been shown[10] to undergo the same colour change at temperatures close to 300 °C. During the process, hydrogen cyanide and ammonia were identified as major gaseous products.

Schwenker and Beck[3] have studied the process for Orlon® in some detail and their results are reproduced in Fig. 17.2. The TG curve, Fig. 17.2(a) shows that

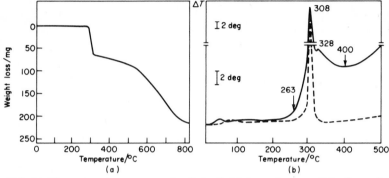

Fig. 17.2. (a) Thermogravimetry curve for Orlon® fabric made from Orlon® acrylic fibre. (b) Differential thermal analysis curves for Orlon® fabric — in air; ––– in nitrogen. (Sample weights 219.2 mg for (a), 150 mg diluted with c. 850 mg alumina for (b)) (Beck and Schwenker, Ref. 3.)

the first loss in weight occurs at around 295 °C and is complete by 320 °C. This process is accompanied by a total weight loss of 25 % and it would appear that the reaction causes a pronounced exothermic heat effect in both air and nitrogen. These authors conclude that the degradation occurs via intermolecular cross-linking with the evolution of HCN. Simultaneously with this effect crystallization of monomer units takes place. Economy et al.[11] have studied the thermal be-haviour of a cross-linked phenol formaldehyde fibre known as Kynol®. Phenolic resins have for many years been used as ablative shields for space re-entry vehicles, because of their ability to withstand temperatures of several thousands of degrees Celsius for a few moments. Kynol® fibres, therefore, would appear to offer a promising possibility for flame-resistant materials. For the DTA studies, samples of fibre (c. 2 mg) were made up of 1–2 mm lengths with a fibre diameter of 20 μm. A heating rate of 20 °C mm^{-1} was used, with a reference sample consisting of glass beads weighing 2 mg. The DTA trace obtained, in air, for Kynol® is shown in Fig. 17.3. The endotherm at about 130 °C is due to the evolution of absorbed water and to the incipient formation of small amounts of peroxide. The exotherm at higher temperatures is to a great extent dependent on the breakdown of these peroxides. The trace obtained for a sample of an acetylated phenolic fibre is also shown on Fig. 17.3. This fibre shows essentially no reaction up to 190 °C and thus acetylation of the Kynol® fibre increases the long term, high-temperature range applications. A further advantage with the acetylated fibre is that it is white and can be dyed a range of colours.

Godfrey[12] has studied the degradation of various rayon fibres containing organophosphorous flame-retardant additives. His results are shown in Fig. 17.4. Curve A is the trace obtained, under a nitrogen atmosphere, for rayon and shows a single endotherm with a peak maximum at 336 °C. Curve B is for the same base material but with the addition of 14% tetrakis (hydroxymethyl) phosphonium chloride and shows an exothermic effect with a peak at 293 °C. Conversely when 8% tris (2,3-dibromopropyl) phosphate is added to rayon, the DTA trace (Curve C) shows a small endotherm with a peak maximum at 232 °C. Curves for the same materials in air reveal a similar pattern, clearly indicating that the addition of these flame retardants in fact lowers the thermal stability of rayon.

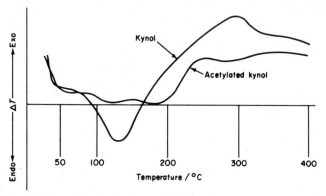

Fig. 17.3. Differential thermal analysis of acetylated Kynol and Kynol. (From Economy *et al.*, Ref. 11.)

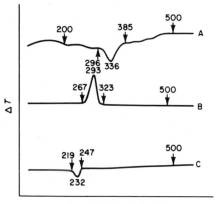

Fig. 17.4. Differential thermal analysis curves for rayon fibre: A, untreated; B, with 14% tetrakis (hydroxymethyl) phosphonium chloride; C, with 8% tris(2,3-dibromopropyl) phosphate. (Stone GS-2 Thermal Analyzer; heating rate 5 °C min⁻¹; flowing N₂ atmosphere). (Godfrey, Ref. 12.)

Finally, DTA has been used[13] to study thiazole polymers for use in highly heat-resistant synthetic fibres. These fibres were aimed especially at military and space applications.

APPLICATION IN FORENSIC SCIENCE

Differential thermal analysis and DSC have proved to be of great importance in the field of forensic science. The ability of this technique to diagnose hair and cloth polymers, using very small samples, enables positive identification or comparison with known materials to be made. This can be used as scientific evidence in criminal investigations. An excellent paper on this subject has been published by Davies.[14]

REFERENCES

1. R. F. Schwenker and P. K. Chatterjee, *Differential Thermal Analysis*, Vol. 2, (R. C. McKenzie, Ed.), Academic Press, New York, 1974, p. 419.
2. L. R. Beck, R. F. Schwenker and R. K. Zuccarello, *Am. Dyest. Rep.* **53**, 30 (1964).
3. L. R. Beck and R. F. Schwenker, *Text. Res. J.* **30**, 624 (1960).
4. A. R. Haly and W. Snaith, *Text. Res. J.* **40**, 898 (1967).
5. M. Ikeda and Y. Mitsuishi, *J. Polym. Sci.* **4**, 283 (1966).
6. H. Kanetsuma and K. Maeda, *J. Chem. Soc. Jpn, Ind. Chem. Sect.* **69**, 1793 (1966).
7. C. M. Conrad, P. Harbrink and A. L. Murphy, *Text. Res. J.* **33**, 784 (1963).
8. L. R. Beck, R. F. Schwenker and R. K. Zuccarello, *Thermal Characteristics of Coated Textile Fabrics*, Final Rept. U.S. Navy Contract N 140 (132), 57422 B, (1960).
9. R. L. Hebert, W. K. Wilson and M. Tryon, *Tappi* **52**, 1183 (1969).
10. R. C. Houtz, *Text. Res. J.* **20**, 786 (1950).
11. J. Economy, F. J. Frechette, G. Y. Lei and L. C. Wohrer, *High Temperature and Flame Resistant Fibres*, Interscience Publishers, Chichester, Sussex, U.K. 1973, p. 81.
12. L. E. A. Godfrey, *Text. Res. J.* **40**, 116 (1970).
13. T. B. Cole, L. G. Picklesimer and W. C. Sheenan, *J. Polym. Sci.* A3, 1443 (1965).
14. G. Davies, *Anal. Chem.* **47**, 318A (1975).

CHAPTER 18

Explosives, Propellants and Pyrotechnics

The use of DTA in this area of study has become more widely accepted in recent years. Much of this increased acceptance is due to the availability of sensitive equipment capable of handling very small weights of sample. In a NASA Report on pyrotechnic hazards classification[1] the following is stated: 'The DTA test is believed to be a useful supplement to pyrotechnic classification tests. It provides for relative rankings of pyrotechnics to each other, to explosives and to other less reactive hazardous materials . . . The results of such rankings appear consistent with other tests in indicating the materials' ease of initiation. It is recommended as a criterion for hazards evaluation purposes.'

GENERAL CONSIDERATIONS

It is possible to distinguish between two classes of materials. On the one hand there are those materials which, although reacting with a highly exothermic heat effect, do not detonate. In explosives terminology these are called 'secondary explosives'. On the other hand there are the materials classified as 'primary explosives', which can be expected to detonate on heating. It is, of course, possible to reduce the sample weights to such a point where there is little risk of instrument damage. If detonation is avoided, for instance by the use of very slow heating rates in an inert atmosphere, the ensuing reaction may well be difficult to characterize.

Certain equipment modifications may well be necessary when dealing with these types of materials. In the case of materials which do not detonate, but merely decompose rapidly, then conventional equipment can be used with small samples. However, where primary explosives are investigated the equipment should be of a suitably robust construction to withstand any detonation. The prime concern should always be towards the safety of the operator and therefore one approach is to use a remote DTA cell. This enables the measuring system to be isolated from the amplification/recording equipment and it can be suitably shielded for protective purposes. An example of such a remote cell, which has been modified[2] for explosives studies, is illustrated in Fig. 18.1. The cell is based on that supplied by Dupont with their series 900 Equipment.

The heating block has been raised and separated from the thermocouple connections by a large insulating disc of asbestos. For the particular work under investigation it was necessary to exclude oxygen from the vicinity of the sample and therefore the cell had to be modified to enable high vacuum ($\approx 10^{-6}$ Torr) to be attained. This necessitated sealing of the electrical connectors, relocation of the gas inlets and modification of the base plate and locking clamps.

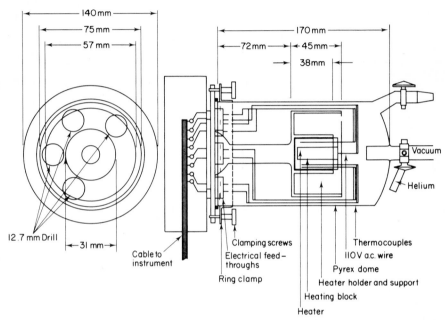

Fig. 18.1. Schematic Diagram of Components in the Modified Remote DTA Cell. (From Graybush *et al.*, Ref. 2.)

The modified cell was mounted onto a brass plate into which four holes were drilled to accept vacuum-tight connectors. A new heating block holder and support was fabricated as a single unit from heavy walled glass and coated with silver. The thermocouple connections were protected by shielding them with glass tubing. Evacuation and gas flow were effected through vacuum stopcocks connected to the top of a pyrex dome; this dome was made vacuum-tight by fitting it to the base via a ring seal and metal collar.

EXPERIMENTAL VARIABLES

Sample preparation

In earlier chapters of this book we have described the effects of particle size, sample preparation, etc. on the shape of a DTA trace and have warned that caution is necessary in interpreting DTA results. The effects of these variables on pyrotechnic studies is well illustrated by the work of Maycock *et al.*[3] on the effect

of sample preparation on ammonium perchlorate. Three methods of crystallization were studied:

(a) from a solution at 60–70 °C cooled to room temperature,
(b) from an aqueous solution cooled to 5 °C,
(c) single crystals grown from a saturated aqueous solution at 45 °C cooled at
the rate of 0.1–0.01 °C per day and crushed to the desired particle size
($\approx$ 100 μm).

The DTA traces obtained (for samples of constant particle size and age) are
shown in Fig. 18.2 and demonstrate clearly the marked differences between the
samples. The differences are thought to be due to trace quantities of perchloric
acid monohydrate diffusing into the ammonium perchlorate crystal lattice, the
effect being more pronounced in the order sample C > B > A.

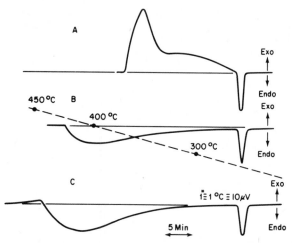

Fig. 18.2. Differential thermal analysis traces for three preparations of NH$_4$ClO$_4$ conducted on
samples in a platinum crucible at a heating rate of 6°C min^{-1} and in a flowing dry He
atmosphere (10 1 h^{-1}). (From Maycock *et al.*, Ref. 3.)

Atmosphere

The gaseous atmosphere around the sample should be carefully controlled in
investigating pseudostable samples. It has been shown,[4] for instance, that small
amounts of water vapour present in the atmosphere can affect the shape of the
DTA trace obtained on the decomposition of ammonium perchlorate. Similarly,
trace amounts of oxygen have a marked effect on the decomposition of lead
azide.[2, 5] Atmosphere problems can generally be overcome by providing efficient
trapping systems in the gas lines for the removal of trace contaminants. In cases
where removal of contaminants in the reaction zone is necessary, it is advisable
to evacuate the complete system prior to filling with a contaminant-free, inert
gas.

EXAMPLES

Graybush *et al.*[2] have studied the decomposition of lead azide using the modified DTA cell described earlier. Sample weights were in the range 0.5–2.5 mg with an equal weight of glass beads as the reference material. The DTA cell was evacuated for several minutes and then backfilled with dried helium. A flow rate of 400 cc min[-1] of helium was used as the sample atmosphere and the heating rate was 10 °C min[-1].

Some results obtained are shown in Fig. 18.3. Figure 18.3(a) shows the DTA trace of a sample which detonated and indicates how little information can be obtained as a result (with the exception of the onset temperature). This result was obtained using the conventional DTA cell and it is assumed that the presence of some residual oxygen was responsible for the detonation. Figure 18.3(b) shows

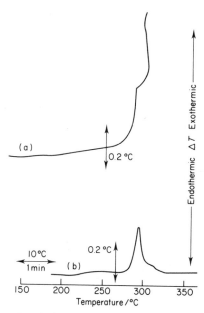

Fig. 18.3. Differential thermal analysis curves for lead azide: (a) sample detonated; (b) sample decomposed to form oxides of lead. (From Graybush *et al.*, Ref. 2.)

the trace obtained on the modified system and illustrates the effectiveness of the sample pre-evacuation techniques employed. Features of importance on the trace are the decomposition exotherm over the temperature range 290–310 °C and the lead fusion endotherm at 318 °C. (The two peaks overlap.)

The choice of helium as the inert gas was made on the necessity of providing a rapid heat distributor in contact with the sample. This was required in order to prevent detonation and to provide controlled decomposition. When the sample size was increased to a point where detonation occurred even in helium, dilution of the sample with glass beads proved effective. Graybush *et al.* also determined the decomposition temperature for potassium dinitrobenzfuran (KDNBF), lead styphnate and mercury fulminate. Hogge[6] has evaluated the feasibility of replacing the atomized aluminium used in high explosive compositions, with low-cost aluminium alloy granules. Differential thermal analysis experiments were performed on a series of mixtures with TNT, ammonium nitrate and Tritonal®. Sample weights varied between 20 mg for the single components to 40 mg for samples diluted with aluminium. A heating rate of 10 °C min⁻¹ was used throughout.

Everett *et al.*[7] have studied the thermal stability of tetracene, using differential scanning calorimetry (DSC). Tetracene is widely used in priming compositions for ammunition and explosive charges for rivets. Samples were sealed in aluminium pans and an atmosphere of helium used throughout the DSC experiment. Heating rates were 0.625, 1.25, 5, 10 and 80 °C min⁻¹. At heating rates of 0.625–10 °C min⁻¹, two distinct exotherms were obtained, whereas at 80 °C min⁻¹ only one was found (Fig. 18.4). At the high heating rate, explosion is believed to have

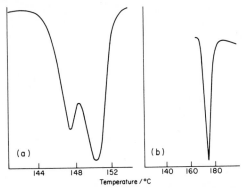

Fig. 18.4a. Differential scanning calorimetry curve for tetracene at 10 °C min⁻¹ (Lot A).
Fig. 18.4b. Differential scanning calorimetry curve for tetracene at 80 °C min⁻¹ (Lot A). (From Everett *et al.*, Ref. 7.)

occurred. An interesting point is that no endothermic melting peak was observed, although tests on a Van der Kamp apparatus (for determining melting points) have indicated that melting does occur prior to decomposition. The authors therefore conclude that the first of the two peaks obtained at 0.625–10 °C min⁻¹ represents the formation of a liquid decomposition product (and nitrogen).

Stability tests were also performed on the material when heated at temperatures of 100 °C, 75 °C and 60 °C for times of 30 min, 24 h and 7 days respectively. After the heat treatment, DSC runs were performed at a heating rate of 1.25 °C min⁻¹. Deterioration of the tetracene (meaning its inability to explode) was evidenced by the disappearance of the second exothermic peak. The authors were unable to give any explanation as to the cause of the disappearance of this peak.

Tetramethylene tetranitramine (HMX) is a material which exists in four conformational forms, but in the crystalline state, the most stable polymorph is the β form. The kinetics of the thermal decomposition of β HMX have been studied by Sinclair and Hondee.[8] Figure 18.5 shows the results obtained for a sample of β-HMX heated, in air, at a rate of rise of temperature of 2 °C min[-1]. The significant features on the trace are the endotherm at 192 °C, attributed to the β–δ phase-change and the exotherm at 276 °C due to violent decomposition. Data

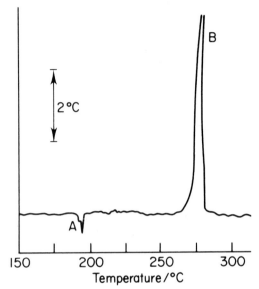

Fig. 18.5. The DTA trace of β-HMX (experimental) at 1 atmosphere, heating rate 2 °C min[-1], shows endotherm at A and exotherm at B. (From Sinclair and Hondee, Ref. 8.)

obtained at different heating rates were analysed by the method of Kissinger (see p. 186). Three distinct slopes, over discrete temperature ranges, of the Kissinger plots were obtained, with activation energies of 184.8, 260 and 220 kJ mol[-1] respectively. Bearing in mind the inadequacies of the Kissinger method as applied to DTA it is difficult to put much physical significance on data obtained from such a highly exothermic process.

Other examples, illustrating the use of DTA in the study of pyrotechnic mixtures may be found in the work of Sulacsik[10, 11] and of Reich.[12, 13] A review of the applications of thermal analysis in general to explosive and solid propellant ingredients has been published by Maycock.[14]

REFERENCES

1. Pyrotechnic Hazards Classification and Evaluation Program, Phase III, Segments 1–4, *Investigation of Sensitivity Test Methods and Procedures for Pyrotechnic Hazards Evaluation and Classification*, April 1971, NASA-CR-122979.

2. R. J. Graybush, F. G. May and A. C. Forsyth, *Status of Thermal Analysis*, NBS Special Publication 338, October 1970, pp. 151–164.
3. J. N. Maycock, L. L. Rouch and V. R. PaiVerneker, *Inorg. Nucl. Chem. Lett.* **4**, 19 (1968).
4. L. L. Bircumshaw and T. R. Phillips, *J. Chem. Soc.* 4741 (1957).
5. J. N. Maycock, L. L. Rouch and V. R. PaiVerneker, *Thermal Analysis*, Vol. 1 (R. F. Schwenker and P. D. Garn, Eds.), Academic Press, New York, 1969, p. 243.
6. W. C. Hogge, *Evaluation of low cost Aluminium Alloy Granules for use in Aluminised explosives*. Report No. NWSY TR 69-2. August 1969, Naval Weapons Station, Yorktown, Virginia, U.S.A.
7. M. E. Everett, M. T. Gurbarg and G. Norwitz, *Thermal and Stability Study of Tetracene Using DSC*, Report No. AD/A-005 163, Frankford Arsenal, U.S.A., December 1974.
8. J. E. Sinclair and W. Hondee, *Jahrestagung* 1971, Institut für Chemie der Treib und Explosivstoffe, p. 57.
9. M. E. Kissinger, *Anal. Chem.* **29**, 1702 (1957).
10. L. Sulacksik, *J. Therm. Anal.* **5**, 33 (1973).
11. L. Sulacksik, *J. Therm. Anal.* **6**, 215 (1974).
12. L. Reich, *Thermochim. Acta* **5**, 433 (1973).
13. L. Reich, *Thermochim. Acta* **8**, 399 (1974).
14. J. N. Maycock, *Thermochim. Acta* **1**, 389 (1970).

CHAPTER 19

Hydraulic Cements

Materials which set and harden due to chemical reactions which occur as a result of mixing with water are termed hydraulic cements. The best known examples are Portland cement which consists mainly of calcium silicates, and alumina cement which contains calcium aluminates. Hydraulic cements will set and harden either under atmospheric conditions, or under water. Although initial hardening occurs quite rapidly, the process continues over a period of several days before full mechanical strength is attained.

The fields in which DTA has been applied to the chemistry of cements can be summarized as follows:

(a) Analysis of the raw materials prior to calcination e.g. to determine the amounts of calcium carbonate and magnesium carbonate present.

(b) A study of the physical and chemical processes which occur as the finely ground raw materials are heated progressively up to about 1500 °C, to form cement clinker.

(c) To study the composition and rate of growth of the hydration products at various periods after setting of the cement.

(d) The effect of accelerators and retarders on the setting characteristics of cements has been studied by DTA.

(e) Differential thermal analysis has now become a standard method for determining the degree of conversion (and consequent loss of strength) in high alumina cements.

For a short and concise description of the chemistry of hydraulic cements, it is suggested that the reader refer to an article by H. F. Taylor,[1] published in the Royal Institute of Chemistry lecture series. A bibliography of the application of DTA to cement chemistry is given by Barta, in Mackenzie's book[2] on DTA.

PORTLAND CEMENT

In the manufacture of Portland cement, a mixture of finely ground limestone, or anhydrite, with clay is suspended in water to form a homogeneous slurry. This slurry is then pumped into the upper end of a long, nearly horizontal, rotating kiln. Burning gases are injected into the lower end of the kiln, so that the cement raw materials are heated to a progressively higher temperature as they pass down the kiln. Water is rapidly lost, followed by decomposition of the calcareous material to give a lime. Reaction between the lime and silicaceous material occurs in the temperature range 1300–1500 °C, at which temperature some 20–30% of the mixture has melted. The reaction product, sintered into rounded lumps of cement clinker, is then cooled and mixed with a controlled amount of gypsum, to retard the setting rate of the cement. Finally the mixture is finely ground, so as to produce ordinary Portland cement with a specific surface of not less than 2250 cm^2 g^{-1}. Rapid-hardening Portland cement is further ground to give a specific surface in excess of 3250 cm^2 g^{-1}.

From the very brief description given above, it will readily be apparent that DTA can be used to simulate quite closely the conditions existing during cement manufacture. The rate of heating can be adjusted so as to approximate to that in the kiln, while the composition of the flowing atmosphere can be matched to that of the burnt heating gases. A filtered portion of the raw material slurry can be used to provide the solid sample required for DTA. Alternatively a range of mixtures covering different materials and compositions can be prepared in the laboratory. A typical DTA curve for a mixture of cement raw materials is shown in Fig. 19.1.

The main constituent of Portland cement is the compound $3CaO.SiO_2$; there is also some β-$2CaO.SiO_2$, together with small amounts of aluminates and ferrites. Hydration and consequent hardening takes place over a period of many days, giving rise to a calcium silicate hydrate of uncertain composition (Taylor[1] suggests $3CaO.2SiO_2.2.5H_2O$) and calcium hydroxide. DTA has been used to study the mechanism of hardening[2, 3] and the effects of various electrolytes in accelerating the hardening process.[4] A method of estimating the degree of hydration of Portland cements, using quantitative DTA has recently been published by Mascolo.[5] This involves grinding the cement with acetone, followed by desiccation for 15 h; the sample is then sieved and the −150 mesh fraction subjected to DTA in static air at 30 °C min^{-1}, using alumina reference material. The sample is first heated to 1000 °C to eliminate thermal effects which had been found to interfere with the 925 °C endotherm, characteristic of unhydrated $3CaO.SiO_2$. This was followed by cooling to 700 °C and then determining the DTA curve up to 1000 °C. The area under the peak occurring in the region of 925 °C is then taken as giving a measure of the amount of unhydrated $3CaO.SiO_2$ present. Initially the apparatus has to be calibrated by using mixtures containing a known amount of $3CaO.SiO_2$ with alumina. Mascolo found that a graph of the 925 °C peak area against $3CaO.SiO_2$ content gave a straight line through the origin. DTA curves for cement after various periods of hydration showed that after one day, hydration was 35%; after seven days it had reached 71% and slightly exceeded 80% after thirty days.

A typical DTA curve for hydrated Portland cement is illustrated in Fig. 19.2 and shows three major endothermic effects. The low-temperature endotherm,

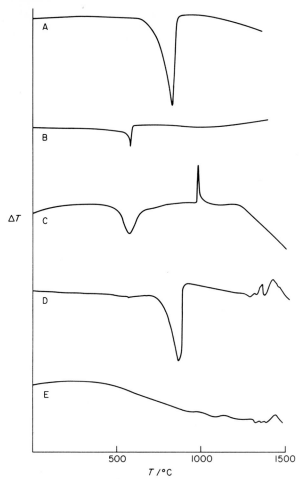

Fig. 19.1. DTA curves for (A) calcium carbonate, (B) silica and (C) kaolin; (D) a mixture of 83.4% calcium carbonate, 8.2% silica and 8.4% kaolin; (E) the residue from D re-heated.

with a peak temperature in the region of 120–165 °C, has been ascribed to loss of water by hydrated calcium silicate gels. This is followed by an endotherm occurring at about 500–540 °C, due to decomposition of calcium hydroxide, which was formed by hydrolysis of some of the calcium silicates originally present. A third endothermic process commences in the vicinity of 800 °C and is associated with complete dehydration of the remaining calcium silicates, together with possible solid–solid phase transitions.

The effect on hardening produced by the various materials added to cement during concrete manufacture has been investigated by Ramachandran,[6] using DTA. Types of additive considered included salts, air entraining agents, pozzolans, colouring agents, etc.

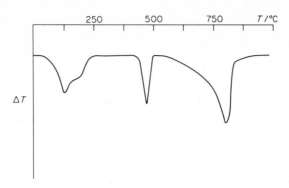

Fig. 19.2. DTA of hydrated Portland cement, water/cement ratio 0.35, after hardening. 87 mg sample, heating rate 10 °C min⁻¹ in flowing nitrogen. (Standata 6-25.)

HIGH ALUMINA CEMENTS

There are basically two types of high alumina cement; a white form for use at high temperatures and a grey form used in the building construction industry. Aluminous cements offer the advantages of rapid hardening, high resistance to chemical attack and superior performance at elevated temperatures. Against this must be set the risk of a progressive and possibly catastrophic loss in strength, which will be discussed in detail later.

Manufacture of aluminous cements involves heating a mixture of bauxite and limestone to a temperature at which fusion occurs, typically about 1500–1600 °C. The furnace is then tapped and the liquid allowed to cool slowly, so as to form slabs consisting mainly of monocalcium aluminate ($CaO.Al_2O_3$). The comparatively high cost of aluminous cements is due very largely to the expense involved in grinding these very hard slabs of cement clinker, in order to produce the required powder. Comments on the use of DTA in studying the manufacture of Portland cement apply generally to aluminous cements, although a higher temperature will be required and the reaction mixture now fuses completely.

Setting and hardening

Aluminous cements harden very much more rapidly than Portland cements, and may gain their maximum attainable strength in as little as 24 h; this contrasts with a period of up to 28 days for Portland cement. It is essential to use a low water:cement ratio to give a strong, low porosity product.[7] The setting reactions may be expressed as follows:

$$CaO.Al_2O_3 \begin{cases} \longrightarrow CaO.Al_2O_3.10H_2O \\ \longrightarrow 2CaO.Al_2O_3.8H_2O + Al_2O_3 \text{ gel} \end{cases}$$

with the upper process predominating. Both hydrated calcium aluminates form hexagonal crystals and are metastable. The decahydrate gives a characteristic

DTA endothermic peak which is rather broad and has a maximum in the region of 110–120 °C.

Conversion and failure

Aften hardening, high-alumina cement slowly degrades, or converts, into decomposition products which have a low inherent strength. This process, which causes a considerable increase in the porosity[8] of the structure, is very slow under dry conditions at normal temperatures. However, at a high relative humidity in a heated building, a disastrous loss in strength may occur well within the designed lifespan of the structure. Such conditions occurred[9] in the building containing the heated swimming pool at the Sir John Cass School, Stepney, London, and led to the failure of concrete, roof-supporting beams in 1974, about eight years after the building was completed.

Unfortunately, there is no indication on the external surface of a high alumina cement structure that conversion has occurred, or to what extent. Within the structure, conversion can cause a characteristic change in colour, from grey to chocolate brown.

The mechanism of the conversion reaction appears[10] to be as shown below.

$$3(CaO.Al_2O_3.10H_2O) \rightarrow 3CaO.Al_2O_3.6H_2O + 2(Al_2O_3.3H_2O) + 18H_2O$$
$$3(2CaO.Al_2O_3.8H_2O) \rightarrow 2(3CaO.Al_2O_3.6H_2O) + Al_2O_3.3H_2O + 9H_2O$$
$$Al_2O_3 \text{ gel} \rightarrow Al_2O_3.3H_2O + nH_2O$$

The conversion products have a cubic structure, whereas the hydrated cement forms hexagonal crystals. Furthermore, the density of the conversion products falls within the range 2.40–2.53 g cm^{-3}, as against 1.72–1.95 g cm^{-3} for the hydrated cement. Since the volume of a cement structure is effectively fixed, it follows that conversion to denser products must be accompanied by a growth in porosity of the solid. This alone could be sufficient to explain the sudden loss in strength which has been found to occur; a 7% increase in porosity can lead to a 50% reduction in strength. However, the mechanism of the conversion process is clearly complex and the complete explanation for the associated loss in strength is still a matter of debate.

Differential thermal analysis of the conversion products shows that tricalcium aluminate hexahydrate gives a strong endothermic peak, with a maximum between 320 °C and 350 °C; alumina trihydrate gives a similar endotherm having a peak maximum between 295 °C and 310 °C.

Sampling and analysis

The procedure set out below is based on that suggested by Midgley and is intended as a guide only. By the time this book is published, it is most probable that various national or international bodies will have drawn up their own test specifications, which should of course be complied with.

To test a suspect concrete beam, first drill a hole to about $\frac{1}{4}$ in depth and remove the dust; then drill a further $\frac{3}{4}$ in and retain all the dust formed. At least three more holes should be drilled in the region under test and the dust samples mixed. Samples should be taken at both ends of a beam and at its mid-point, together with any other points at which the beam is supported. It is important not to allow the drill to overheat, in case this has an effect on the calculated percentage conversion.

The samples are then submitted to DTA using a static air atmosphere and a 10 °C min⁻¹ rate of rise of temperature, with sintered alumina as the reference material. A maximum temperature of about 400 °C will be required. However, in order to confirm that the samples do not contain any hydrated Portland cement, it is necessary to raise the temperature to about 600 °C. The presence of Portland cement will then be indicated by the appearance of an endotherm at about 520 °C, due to calcium hydroxide.

The percentage conversion is given by:

$$\frac{\text{mass of } Al_2O_3.3H_2O \times 100}{\text{mass of } Al_2O_3.3H_2O + \text{mass of } CaO.Al_2O_3.10H_2O}$$

If it is assumed that the mass of a substance present is proportional to DTA peak height, then

Conversion (%) =

$$\frac{B \times \text{peak height } Al_2O_3.3H_2O \times 100}{B \times \text{peak height } Al_2O_3.3H_2O + C \times \text{peak height } CaOAl_2O_3.10H_2O}$$

The constants B and C must be determined using a standard material, whose composition is known. Fortunately, it appears from the results of several investigators that for all practical purposes it is safe to assume $B = C = 1.0$. Since the conversion product $3CaO.Al_2O_3.6H_2O$ may also show up on the DTA trace, Midgley proposed that only the height of the lower temperature peak, occurring within the range 295–350 °C, should be taken as a measure of the mass of conversion products. An accuracy of $\pm 5\%$ is claimed for this method. Some typical DTA curves for partially converted high alumina cements are shown in Fig. 19.3. For very high conversions, say above 90%, it becomes increasingly difficult to make an accurate calculation, because the DTA endotherm due to the unconverted decahydrate grows progressively broader and more indistinct. This is well illustrated by comparing curve A with curve B in Fig. 19.3.

The degree of reproducibility obtainable can be judged from Table 19.1, which records the percentage conversion of samples supplied by the Building Research Station. The figures in the left-hand column were mean values determined by BRS, using DTA, while those on the right were obtained at Portsmouth Polytechnic, using a Standata 6-25 differential thermal analyser.

The presence of Portland cement will be demonstrated by the appearance of an endothermic peak in the region of 520 °C, due to decomposition of calcium hydroxide, provided that the sample is heated to a sufficiently high temperature.

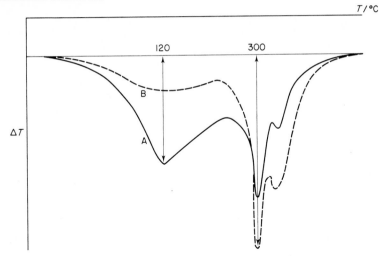

Fig. 19.3. DTA curves for partially converted high alumina cement. — 55% and --- 88% conversion. (Static air, 10°C min⁻¹, Standata 6-25.)

While in theory it would be desirable to measure peak areas rather than peak heights, the way in which the peaks overlap makes this impossible.

Differential thermal analysis, together with other techniques, has been used to investigate[11] the effect of varying the water/solids ratio and changes in temperature on the physical properties of alumina cements, after hardening.

TABLE 19.1

Percentage conversion of samples of hydrated high-alumina cements

Mean values found by Building Research Station	Values found at Portsmouth Polytechnic, using a Standata 6-25
55	56.8, 56.7
70	66.8, 72.4
88	83.4, 84.9

GYPSUM PLASTERS

The dehydration of gypsum, $CaSO_4.2H_2O$ is used to prepare a variety of plaster type of products, whose compositions range from $CaSO_4.\frac{1}{2}H_2O$ down to forms of anhydrite containing very little chemically bound water. This process can readily be followed by DTA, since the mechanism of dehydration passes through a number of clearly defined, consecutive stages as the temperature is increased.

$$CaSO_4.2H_2O \rightarrow CaSO_4.\tfrac{1}{2}H_2O \rightarrow CaSO_4.\epsilon H_2O \rightarrow CaSO_4$$
$$\text{(hexagonal)} \quad \text{(orthorhombic)}$$

Dehydration of gypsum gives rise to an endothermic peak, having an onset temperature in the region 112–115 °C and a peak temperature of 140–150 °C.

Two forms, α and β, of the semihydrate exist; decomposition gives rise to a second endotherm overlapping the first, but having a clearly defined, separate peak. It is possible to distinguish between the α and β forms using DTA; characteristic curves are given in a paper by Murat.[12] The dehydration product still contians a small amount of chemically bound water, $\epsilon \leqslant 0.15$, which is subsequently evolved in an exothermic step occurring at between 300 and 400 °C in the β-semihydrate. The α-semihydrate shows an exothermic peak at a much lower temperature, adjacent to the endotherm due to decomposition of the semihydrate itself (see Fig. 19.4).

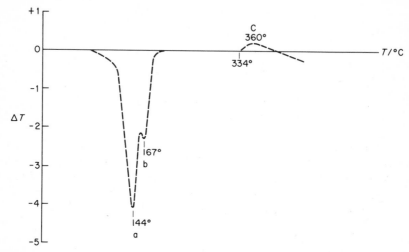

Fig, 19.4. DTA curve for $CaSO_4 . 2H_2O$, showing (a) $-1.5H_2O$ (b) $-0.5H_2O$ and (c) the hexagonal–orthorhombic phase transition (28 mg sample in N_2 atmosphere, flowing; heating rate 10°C min^{-1}).

Hexagonal $CaSO_4 . \epsilon H_2O$ readily re-forms the semihydrate in the presence of water vapour, while the higher temperature orthorhombic form of anhydrite does not. Khalil and Gad[13] give DTA curves for the dehydration of gypsum heated at 90 °C and 160 °C for periods of up to four days. They also give DTA curves for rehydration of the product, obtained after 30 h at 160 °C, then allowed to stand under atmospheric conditions for various periods of up to 90 days.

Satava[14] has studied the effect of varying the ambient water vapour pressure, from 0.025 to 2 atmospheres, on the dehydration of gypsum. His results showed that the peak onset temperature increased with increasing water vapour pressure and that peak separation improved. Measurements have also been made in the presence of liquid water at high pressures,[15] corresponding to temperatures of up to 350 °C.

REFERENCES

1. H. F. W. Taylor, *The Chemistry of Cements*, Royal Institute of Chemistry Lecture Series 1966, No. 2.

2. R. C. Mackenzie, *Differential Thermal Analysis*, Vol. 2, Academic Press, London, 1972.
3. T. L. Webb and J. E. Kruger, *Analysis of Calcareous Materials*, Society of Chemical Industry, Monograph No. 18, 1964, pp. 419–439.
4. J. Teoreanu and M. Muntean, *Rev. Roum. Chim.* **19**, 371 (1974).
5. G. Mascolo, *J. Thermal Anal.* **8**, 69 (1975).
6. V. S. Ramachandran, *Thermochim. Acta* **4**, 343 (1972).
7. H. G. Midgley, *Trans. Br. Ceram. Soc.* **66**, 161 (1967).
8. H. G. Midgley and K. Pettifer, *Trans. Br. Ceram. Soc.* **71**, 55 (1972).
9. S. C. C. Bate, *Report on the failure of roof beams at Sir John Cass's Foundation and Red Coat Church of England Secondary School, Stepney*, Building Research Establishment Publication, CP 58/74.
10. H. G. Midgley, *The Determination of the Degree of Conversion of High Alumina Cements*, Building Research Station Publication (1974).
11. V. S. Ramachandran and R. F. Feldman, *J. Appl. Chem. Biotechnol.* **23**, 625 (1973).
12. M. Murat, *J. Thermal Anal.* **3**, 259 (1971).
13. A. A. Khalil and G. M. Gad, *J. Appl. Chem. Biotechnol.* **22**, 697 (1972).
14. V. Satava and B. Zbuzek, *Silikaty* **15**, 127 (1971).
15. V. Satava, *Silikaty* **15**, 1 (1971).

CHAPTER 20

Glass and Ceramics

INTRODUCTION

Ceramic products cover a wide range of time from the first man-made products (such as pottery and bricks from around 3000 B.C.) to some electrical and magnetic materials which were unknown until recent years. Initially the term ceramics was applied only to clay products, thus excluding glass, cements, enamels and abrasive. However, a broader concept of ceramics which includes these above-mentioned materials is gradually being accepted.

A great many ceramic materials contain silica. In fact, until recently ceramic technology could still be considered as high-temperature silicate chemistry. However, in view of the large number of more recently discovered ceramic materials which have no silicon in them, this definition is now obsolete.

Ceramics have been associated with developments in the field of electricity since very early days. For instance, glass was an important material in the Leyden jar and a glass rod was first used to generate static electricity. Porcelain was one of the materials used as an insulator with the advent of electric power around 1900, while porcelain insulators were first used for overhead telegraph lines in 1842.

Ceramic materials are undergoing extensive development for the electrical industry since they can be used in magnets, condensers, resistors, transformers, memory devices, transducers, capacitors and insulators.

The first ceramic material used as a dielectric was titania. Because of its high dielectric constant and negative temperature coefficient, titania is widely used in the manufacture of capacitors. Needless to say, the work on titania led to the discovery of other materials with even higher dielectric constants. Typical of these is barium titanate, $BaTiO_3$, which has a dielectric constant of about 2000 (20 times that of TiO_2). This material also has the unusual property, in its ceramic form, that it can be made piezoelectric (i.e. converts mechanical deformation into electrical voltage and vice versa).

The development of magnetic ceramic materials started about 1930. These materials are very useful at high frequencies because their high electrical resistivity results in low eddy current losses. Examples of magnetic ceramic materials are the ferrites, of general formula $MO.Fe_2O_3$, where M may be such divalent ions as Cu, Ni, Co, Fe, Zn, Mg or Mn, alone or in combination.

This general introduction therefore illustrates the vast range of materials covered by the general term 'ceramics'. In this chapter we shall restrict ourselves solely to some examples of the way DTA has been used to investigate glass-forming reactions and to study some electrical glass ceramics. Cement materials have been described earlier in Chapter 19 and clays are the subject of an excellent book by Mackenzie.[1]

EXPERIMENTAL STUDIES

Glass may be defined as a solid material without long-range order in its structure. It is isotropic and gradually becomes liquid as the temperature is raised. This is, of course, in sharp contrast to a crystalline substance which has a clearly defined melting point. Mediaeval window glass was a potassium–calcium silicate, the alkali being obtained from the ashes of plants, whereas modern glass is a sodium–calcium silicate. In practice, ground quartz, flint or clean sand (free from iron), mixed with potassium carbonate or sodium carbonate and the other ingredients are fritted together in ovens or furnaces of various designs, by which means the silica of the quartz or sand enters into combination with the bases to form glass. Common glass has a composition which may be represented by the formula $Na_2O.CaO.6SiO_2$.

Differential thermal analysis curves obtained with glass samples are characteristic of the material under test and, in common with most other applications, changes in composition have a marked effect on the shape of the DTA trace. Differential thermal analysis can often provide information as to the correct temperature to be used in the heat treatment process. Figure 20.1 illustrates a DTA trace obtained[2] on heating a barium titanate glass sample. It has been found that heating the sample to a temperature midway between the ferroelectric phase crystallization exotherm and the first melting endotherm is optimum for this type of glass ceramic.

Thomasson and Wilburn[3-5] have published a series of papers which elegantly demonstrate the application of DTA (and TG) to the study of reactions between glass-making materials. Let us take, as an example, the system SiO_2–Na_2CO_3–NaF; as previously discussed for a three-component system, it is of course necessary to consider the three single materials and the three possible two-component systems. Figure 20.2 is the DTA trace obtained for a mixture of 85 mol parts SiO_2, 14 mol parts Na_2O (added as Na_2CO_3) and 2 mol parts NaF. All the ingredients were carefully weighed and mixed beforehand to ensure a uniform sample. Particle size was carefully controlled and in this particular case all particles passing through a BS 300 mesh sieve were used. A heating rate of $10\,°C\ mm^{-1}$ was used for both the TG and DTA experiments.

The DTA curve shows endothermic peaks at 350, 470, 540, 575, 680 and 760 °C and an exotherm at 710 °C. The DTG curve shows a broad peak with an onset temperature of 400 °C. Two-peak maxima are apparent; the first at 630 °C with a second and larger one at 690 °C.

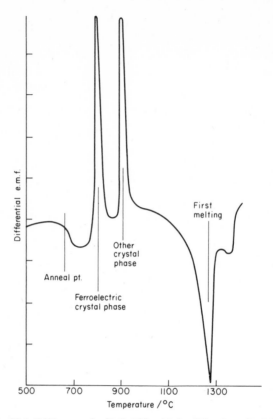

Fig. 20.1. DTA curve for BaO–TiO₂–Al₂O₃–SiO₂ glass. (Ref. 2.)

At this point it should be very clear why it was necessary to look at the one-component and two-component mixtures. Without the information provided by these experiments it would be impossible to attempt an explanation of the DTA trace for the three-component system, even with the TG data. With this preliminary data it is possible to characterize the DTA peaks at 350 °C and 470 °C as being due to phase changes in the sodium carbonate and that at 540 °C as being due to a chemical reaction which results in the formation of sodium meta-silicate. The peak at 575 °C is caused by the heat required for the α–β quartz inversion and the eutectic melting of sodium chloride, sodium fluoride and sodium carbonate.

The first DTG peak occurs at 630 °C and does not correspond to any peak on the DTA curve. This is not really surprising when one considers the complexity of the chemical processes which occur in this temperature range. The DTA peak at 680 °C is accompanied by a loss in sample weight and would appear to be associated with a reaction between sodium carbonate and silica. This reaction is undoubtedly enhanced by the presence of molten sodium carbonate.

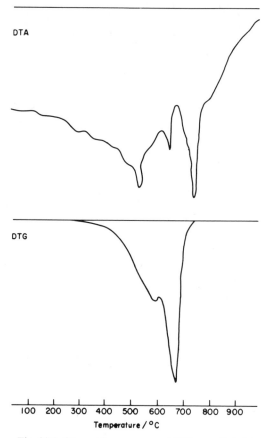

DTA

DTG

100 200 300 400 500 600 700 800 900
Temperature / °C

Fig. 20.2. (From Thomasson and Wilburn, Ref. 4.)

The exothermic peak at 710 °C is due to the chemical reaction between sodium carbonate and silica to form sodium silicate, a process which was shown to occur for the sodium carbonate–silica two-component system. Similarly it is shown that the DTA peak at 760 °C is due to the eutectic melting of sodium disilicate and silica. A summary of the thermal effects obtained during heating is given in Table 20.1.

A similar study to the one described above has been published by Ott and McLaren.[6] They investigated the melting behaviour of a sodium–lead–silicate glass batch. As before, all individual components were studied as well as all possible combinations of two components. All simultaneous TG–DTA experiments were performed in a flowing air atmosphere (dried with molecular sieve) with alumina as reference material. A heating rate of 8 °C mm^{-1} was used. X-ray diffraction enabled the identity of the reaction products to be determined.

Figure 20.3 illustrates the TG and DTA curves obtained on heating a sample mixture consisting of 60% silica, 33.6% litharge and 6.4% soda ash, up to a

TABLE 20.1

Identification of DTA peaks for the system 84 mol parts SiO_2–14 mol parts Na_2CO_3–2 mol parts NaF.

Peak temperature/°C	Thermal effect	Identification
350	Endotherm	Crystalline inversion in Na_2CO_3
470	Endotherm	
540	Endotherm	Formation of Na_2O, SiO_2
575	Endotherm	α–β quartz inversion and NaF–NaCl–Na_2CO_3 eutectic
680	Endotherm	NaF–Na_2CO_3 eutectic and chemical reaction SiO_2–Na_2CO_3
710	Exotherm	Formation of $Na_2O.2SiO_2$
760	Endotherm	Eutectic melting SiO_2–Na_2O $2SiO_2$ and glass formation

temperature of 1000 °C. At 605 °C a DTA peak is observed; this can be attributed to the formation of a sodium lead silicate. The decomposition reaction proceeds slowly over the temperature range 650–750 °C and then more rapidly to a conclusion by a temperature of 850 °C. After 850 °C a liquid phase exists which gradually assimilates the remaining silica as the temperature increases.

The degradation of the silicon–oxygen bond is necessary to form a glass. Near to the liquidus temperature this rate of assimilation is low. The rate increases, however, as the temperature increases. In practice, the practical glass melting

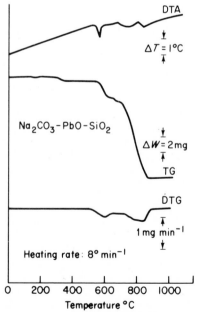

Fig. 20.3. Differential thermal analysis curve for a lead–glass mix. (From Ott and McLaren, Ref. 6.)

temperature usually exceeds the liquidus temperature of the finished glass by between 300 °C and 500 °C; this higher temperature is necessary to provide sufficient energy to assimilate the silica.

Matusita et al.[7] have studied the crystallization of Li_2O–SiO_2 glasses. These glasses are bases for typical commercial glass ceramics. It is necessary for the production of glass ceramics that the glass should crystallize at temperatures at which its viscosity is still high enough to retain the shape of a formed article. The compositions of the glasses used were $33.3Li_2O.66.7SiO_2.3RO_n$, and $25Li_2O.75SiO_2$, RO_n is in moles, where R = Na, K, Cs, Mg, Ca, Sr, Ba, B, Al, In, Ge, Ti, Zr, P or V. The bulk glass was used as the sample instead of a mixture of the ingredients in order to minimize the effects of surface crystallization. To get the sample into the sample holder, the molten glass (1.3 g) was poured in, re-melted at 1400 °C for about ten minutes and then allowed to cool in air. A heating rate of 10 °C min^{-1} was used in all the DTA experiments.

Figure 20.4 shows the DTA curves obtained for $33.3Li_2O.66.7SiO_2.3RO_n$ glasses where R = Mg, Ca, Sr or Ba. The large exotherm at c. 700 °C was found to be due to the precipitation of lithium disilicate ($Li_2O.2SiO_2$) crystals. The endothermic peaks at about 1000 °C for all the glasses studied can be attributed

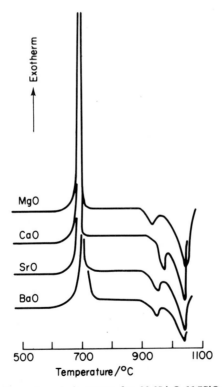

Fig. 20.4. Differential thermal analysis curves for $33.3Li_2O.66.7SiO_2.3RO_n$ glasses. (From Matusita et al., Ref. 7.)

to the fusion of the lithium disilicate. The authors also demonstrated that a relationship exists between the ease of crystallization of the lithium disilicate (as measured by the temperature of the exotherm) and the temperature at which the viscosity reaches 10^{10} P.

These same authors have also described[8] a method for the determination of the activiation energy for crystal growth in the same glass from DTA measurements. From theoretical considerations, the following equation was obtained:

$$\ln\left[C_\mathrm{P}\frac{d(\varDelta T)}{dt}+K\varDelta T\right] = -\frac{E_\mathrm{D}}{RT}+\text{constant} \tag{1}$$

where C_P is the heat capacity of the sample and sample holder, K is the heat transfer coefficient, $\varDelta T$ is the temperature difference between sample and reference at time t, and E_D is an energy term indicative of the activation energy for crystal growth.

The heat transfer coefficient is obtained from the relationship

$$\varDelta H = KA \tag{2}$$

which relates total heat evolved or absorbed to DTA peak area. For the determination of K, DTA traces of the melting of sodium chloride and the solid transition of lithium sulphate were determined for five different sample weights and two different heating rates. A plot of $\varDelta H$ versus A gave a value for K of 0.30 cal mm^{-1} °C^{-1}.

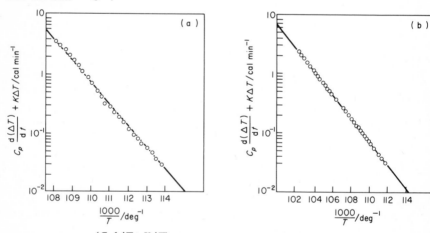

Fig. 20.5. The term $\dfrac{(C_\mathrm{P}d\varDelta T+K\varDelta T)}{dT}$ plotted against reciprocal of absolute temperature for the heating rate of 5°C min^{-1} (a) Li$_2$O.2SiO$_2$ glass, (b) 33.3Li$_2$O.66.7SiO.3TiO$_2$ glass. (From K. Matusita et al., Ref. 8.)

Figure 20.5 illustrates the results obtained for the plot of $\ln[C_\mathrm{P}d(\varDelta T)/dt+K\varDelta T]$ versus I/T for the crystallization of an Li$_2$O.2SiO$_2$ glass and a 33.3Li$_2$O.66.7SiO$_2$.3TiO$_2$ glass at a heating rate of 5 °C min^{-1}.

As can be seen, a good straight-line fit to the experimental data is obtained. A similar result was obtained for both glasses at heating rates of 1, 2, 10, 20 and 40 °C min^{-1}.

The results indicate that the activation energy for crystal growth is $1/3\, E_D$ when bulk nucleation is dominant and is equal to E_D when surface nucleation is dominant, and thereby confirm the theoretical approach.

Shimakawa et al.[9] have used DTA to investigate differences between the thermal properties of 'threshold type' and 'memory type' chalcogenide glass semiconductors. The composition of the materials used were $Te_{65}Se_{15}Ge_{20}$ and $Te_{45}Se_{50}Sb_5$ (atomic %). Figure 20.6 is a DTA trace obtained on heating at a rate of rise of temperature of $10\,°C\,min^{-1}$, with subsequent cooling at $10\,°C\,min^{-1}$ for the $Te_{45}Se_{50}Sb_5$ glass. On heating a glass transition (T_g) at $72\,°C$, an exotherm at $115\,°C$ (T_{ct}) and an endotherm at $335\,°C$ (T_m) are obtained. At T_{ct},

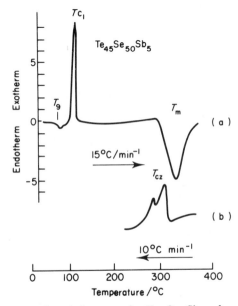

Fig. 20.6. Differential thermal analysis curves for $Te_{45}Se_{50}Sb_5$, where the heating rate was $15°C\,min^{-1}$ and the cooling rate was $10°C\,min^{-1}$. (From Shimakawa et al., Ref. 9.)

X-ray powder studies show that an Sb crystallite is present. When the sample is cooled from a temperature just below T_m two exotherms are obtained around $300\,°C$. The authors ascribe this effect as being related to the high conductive memory filament of the material.

Differential thermal analysis has been used by Hasegawa et al.[10] to study phase relations and crystallization of glass in the system PbO–GeO_2. In the preparation of starting materials having the molar composition ranging from $58PbO.42GeO_2$ to $65PbO.35GeO_2$ in the system $PbO.GeO_2$, the required amounts of lead oxide (99.9 %) and germanium dioxide (99.999 %) were carefully mixed. About 50 g of the mixture was preheated at $550\,°C$ for 20 h followed by melting at $1000\,°C$ in a platinum crucible. After melting, the melt was poured into a mould and formed into a plate of about 1 mm thickness. Differential thermal analysis samples (20 mg) were taken from the plate glass in the form of coarse-grained powders. The DTA traces obtained for the series of compositions studied are given in Fig. 20.7 (PG-62 refers to 62 mol% Pb etc.). If the DTA trace for the PG 62.5 glass (trace (e) in Fig. 20.7) is considered there are three exotherms and one endotherm apparent. The first exothermal step at 334 °C corresponds to the glass transition and the two sharp exotherms in the tempera-

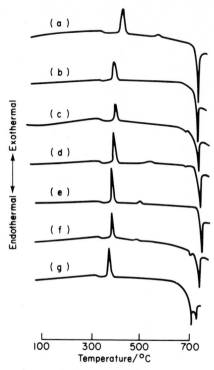

Fig. 20.7. Differential thermal analysis traces of glasses, (a) PG-58 (b) PG-60 (c) PG 61 (d) PG 62 (e) PG 62.5 (f) PG 63, and (g) PG 65 at the heating rate of 10°C min^{-1}. (From Hasegawa *et al.*, Ref. 10.)

ture ranges 372–403 °C and 482–514 °C correspond to crystallization and recrystallization respectively. From the results of X-ray diffraction analyses on these two phases, it would appear that the crystalline phase present in the temperature range 403–480 °C is $Pb_3Ge_2O_7$, whereas at temperatures above 483 °C the product is $Pb_5Ge_3O_{11}$. The third exotherm is described as being due to the chemical reaction between $Pb_3Ge_2O_7$ and residual glass matrix to form $Pb_5Ge_3O_{11}$; this material melts at 737 °C. By this means the authors were able to elucidate the phase diagram, over a restricted composition range, for the system PbO–GeO_2.

REFERENCES

1. R. C. Mackenzie, *Differential Thermal Investigation of Clays*, Mineralogical Society, London, 1957.
2. British Patent No. 905 253.
3. C. V. Thomasson and F. W. Wilburn, *J. Soc. Glass Technol.* **42**, 158T (1958).
4. C. V. Thomasson and F. W. Wilburn, *Phys. Chem. Glasses* **1**, 52 (1960).
5. C. V. Thomasson and F. W. Wilburn, *Phys. Chem. Glasses* **2**, 126 (1961).
6. W. R. Ott and M. G. McLaren, *Thermal Analysis, Proceedings of the Second International Conference for Thermal Analysis*, Vol. 2, Academic Press, New York, 1969, p. 1329.
7. K. Matusita, S. Sakka, T. Maki and M. Tashiro, *J. Mater. Sci.* **10**, 94 (1975).
8. K. Matusita, S. Sakka and Y. Matusui, *J. Mater. Sci.* **10**, 961 (1975).
9. K. Shimakawa, Y. Inagaki and T. Arizumi, *Jpn J. Appl. Phys.* **11**, 1319 (1972).
10. H. Hasegawa, M. Shimada and M. Koizumi, *J. Mater. Sci.* **8**, 1725 (1973).

CHAPTER 21

Coordination Compounds

INTRODUCTION

Very early on in the development of synthetic inorganic chemistry, compounds were prepared which, at first sight, appeared to be a stoichiometric mixture of two independent compounds. Typical of such salts are the alums, for instance the compound $Al_2(SO_4)_3K_2SO_4.24H_2O$. A second group of compounds were also isolated, but in this case one of the salts was replaced by a neutral molecule such as ammonia. Classic examples of this type of compounds are the ammines $NiCl_26NH_3$ and $CuSO_4.4NH_3$.

Studies into these compounds lead to the Coordination Theory and to the concept of coordination number. This is the total number of anions, or molecules, which may be directly associated with a metal cation. Over the last few years, intensive research into these coordination compounds has revealed an enormous range of products and much effort has been expended on structure determinations and other properties. Recently workers have turned to the use of DTA as a complementary technique for these investigations. Because of the many papers which have appeared on this subject it is not practicable to go into large numbers of examples. We have therefore selected some typical examples which should illustrate the application of DTA to this subject. The reader is also advised to consult the texts by Wendtlandt and Smith[1] and Ashcroft and Mortimer[2] for a more detailed coverage. For an excellent basic text on Coordination Complexes, the book by Cotton and Wilkinson[3] should be used.

EXPERIMENTAL TECHNIQUE

There are very few precautions to be taken specifically for these compounds and therefore the basic experimental techniques described earlier are applicable. In the case of compounds made from precious metals, a particular problem arises because of the high cost of such materials. Obviously in this case as small samples as practicable must be used. With the newer types of equipment sample weights of the order of 1 mg can be used, preferably with an empty reference crucible. If the temperature range of investigation is low (up to $\sim 400\ °C$) then aluminium or glass can be used as the crucible material. For phase transition studies many workers prefer to seal the sample into an aluminium crucible in order to eliminate atmosphere effects. When a higher temperature range is required, a problem arises in using platinum crucibles for studies on compounds containing metals such as platinum, palladium, etc. It is frequently found that the end product of the reaction is the metal itself or, when a halogen is present, an active halide may be formed; this tends to cause reaction with the crucible and can lead to a gradual increase in weight (and possibly in value) of the sample crucible. Such problems can be overcome by using one of the dilution techniques referred to earlier (see p. 26).

Some of the earlier work on decomposition studies of these materials utilized DSC rather than conventional DTA, especially where heats of reaction were being determined. Other workers then assumed that only DSC could be used for such work. It should be noted that a properly calibrated DTA equipment can yield just as meaningful data.

EXAMPLES

Allen et al.[4, 5] have elegantly demonstrated the use of DTA in the observation of phase transitions in square planar complexes of palladium and platinum with phosphine ligands. These complexes of the type MX_2L_2 (where $M = Pt(II)$ or Pd(II), X = halide, L = neutral ligand) are known to undergo cis to trans isomerization both in solution and in the solid state, under the action of heat. Examples are also known for the trans to cis isomerization, but these are relatively rare.

Figure 21.1 illustrates the DTA curves obtained for compounds of the form $PdCl_2 (PhPR_2)_2$ when R = methyl or ethyl. Samples of 5 mg were used in a flowing nitrogen atmosphere with a rate of rise of temperature of 10 °C min^{-1}. Thermogravimetry experiments under similar experimental conditions over the temperature range, room temperature to 250 °C, showed that no weight loss occurred and therefore any thermal effect on the DTA trace can be attributed to phase transitions. Consider firstly Fig. 21.1(a). The sharp exotherm ($T = 172$ °C) corresponds to the trans–cis isomerization and is followed by the endothermic melting peak for the cis isomer at 197 °C. Confirmation of the trans–cis isomerization was obtained by far-infrared spectroscopy.

A sample of the cis isomer, when heated under the same conditions, gave the DTA trace illustrated in Fig. 21.1(b). As can be seen, two effects are apparent; a sharp endotherm at $T = 158$ °C and a further endotherm at 196 °C. This latter peak can be attributed to the melting of the cis isomer. Observation of the sample under a hot stage microscope showed that no melting occurred near to a temperature of 158 °C but that thermal agitation of the sample was apparent. The endotherm at 158 °C, therefore, must be due to some form of solid-phase transition. Cooling of the sample and reheating eliminated this effect and an exactly similar result was obtained on cooling and reheating a sample of the trans isomer.

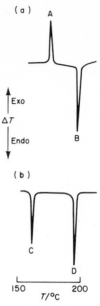

Fig. 21.1. Differential thermal analysis curves for various phosphine complexes of palladium, (a) *trans*-PdCl₂(PhPMe₂)₂; (b) *cis*-PdCl₂(PhPMe₂)₂. (From Allen *et al.*, Ref. 5.)

Cis–trans isomerization is reported by these workers for the corresponding diethylphosphine compounds. Heats of isomerization were calculated by calibration of the equipment using the melting of indium. Similar phase-transition studies on compounds of the form [CrCl₂(en)₂] Cl are reported by Sato and Hatakeyama.[6]

Tetahata *et al.*[7] have studied the effect of temperature upon the solid state racemization of phenanthroline or dipyridyl-iron (II) and nickel (II) perchlorates. It was shown that racemization of these complexes occurred after the dehydration process and the authors report the activation energies for the racemization.

An example of the combined use of TG and DTA is given by Amigo *et al.*[8] in their study on the thermal behaviour of [CoCO₃(NH₃)₄]₂SO₄.3H₂O. The DTA curve for a sample (*c.* 150 mg) heated at a rate of rise of temperature of 10 °C min⁻¹ in an atmosphere of air is shown in Fig. 21.2. The corresponding TG curve (5° min⁻¹, 505 mg) is illustrated in Fig. 21.3. As indicated on the TG curve, a weight loss of 9.7% occurs over the temperature range 100–190 °C. This weight loss agrees reasonably well with that calculated for the loss of the three molecules of water of crystallization (10.3%). The process is accompanied by two endotherms on the DTA curve, with peak maxima at temperatures of 120 °C and 155 °C. Between 240 and 380 °C a further loss in weight of 40.7% is found and is accompanied by a large endothermic effect and an associated small exotherm. Infrared spectra of the sample removed from the DTA equipment at this point showed an absence of absorption bands due to NH₃ or CO₂. A theoretical value

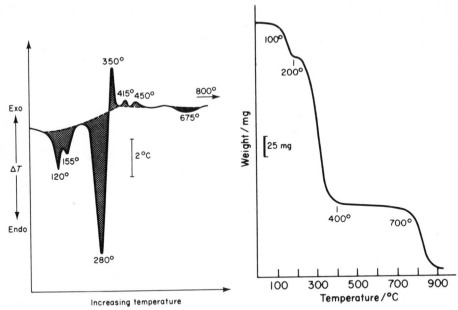

Fig. 21.2. Differential thermal analysis curve of $[CoCO_3(NH_3)_4]_2SO_4 \cdot 3H_2O$ in air atmosphere; mass of the sample; 150 mg; inert reference material: α-Al_2O_3; thermocouples: chromel/alumel; sample holder: Ni: rate of heating: 10°C min^{-1}. (From Amigo et al., Ref. 8.)

Fig. 21.3. Mass-loss curve of $[CoCO_3(NH_3)_4]_2SO4 \cdot 3H_2O$ in air atmosphere; 505 mg; rate of heating: 5°C min^{-1}. (From Amigo et al., Ref. 8.)

for complete loss of these species would be 42.7%. The final product obtained at 800 °C is shown by X-ray diffraction to be Co_3O_4. Further studies on cobalt amine complexes are reported by Halmos and Wendtlandt.[9] Verdonk[10] has studied the synthesis and thermal decomposition of trioxalato-aluminate hydrate and is another good example of the combined use of TG and DTA.

Savant and Patel[11] have studied the thermal decomposition of some diphenyl sulphoxide (DPSO) complexes of thorium (IV). $[ThCl_4(DPSO)_4]$, $[ThBr_4(DPSO)_4]$ and $[Th(NO_3)_4(DPSO)_3]$ decompose endothermically both in air and nitrogen, whereas $[Th(DPSO)_6(ClO_4)]$ and $[Th(NCS)_4(DPSO)_4]$ decompose exothermically, the former decomposing extremely violently. This violent reaction is presumably due to the high oxygen content of the perchlorate. Carbon is formed during the decomposition of all of the complexes and, in air, leads to an exotherm on the DTA trace at a temperature of approximately 550 °C. In nitrogen, this reaction does not occur, but otherwise the DTA traces are unaffected by changes in atmosphere.

Steger[12] has determined what influence, if any, atmospheric parameters have on the decomposition of the metal complexes of 1,5-bis (2 mercapto-ethylthio) pentane. The metals studied were Fe(III), Co(II), Cu(I) and Zn(II), and the TG and DTA traces obtained are shown in Fig. 21.4. In each case, it is clearly

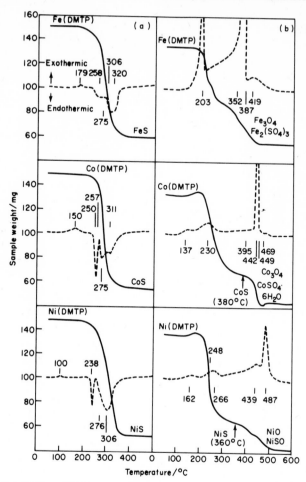

Fig. 21.4. Combined TG and DTA curves for various coordination complexes. (From Steger *et al.*, Ref. 12.)

demonstrated that when the decomposition process is effected under nitrogen the corresponding sulphide is formed as the final product, whereas in air mixed oxides and sulphates result. There is strong evidence to suggest that, certainly in the case of the nickel and cobalt complexes, the sulphide is formed as an intermediate product but that it undergoes subsequent oxidation as the reaction proceeds. Steger also demonstrates that a correlation exists between the decomposition temperature under nitrogen and the reported stabilities of the complexes of the metal ions with sulphur–donor ligands.

This, then, has been a very brief description of the sort of applications to which DTA can be put when studying coordination compounds. Although only a few examples have been given, the reader should be aware that decomposition of these materials is frequently a complicated process and that DTA cannot be used alone. Thermogravimetry, X-ray, i.r. and hot stage microscopy are techniques commonly employed in order to elucidate the reaction mechanism; in many cases direct analysis of the gaseous products may well prove necessary.

REFERENCES

1. W. W. Wendtlandt and J. P. Smith, *Thermal Properties of Transition Metal Ammine Complexes*, Elsevier, Amsterdam, 1967.
2. S. J. Ashcroft and C. T. Mortimer, *Thermochemistry of Transition Metal Complexes*, Academic Press, London, 1970.
3. F. A. Cotton and G. Wilkinson, *Advanced Inorganic Chemistry. A Comprehensive Text*, Interscience Publishers, Chichester, Sussex, U.K. 1972.
4. E. A. Allen, N. P. Johnson, D. T. Rosevear and W. Wilkinson, *Chem. Comm.* 171 (1971).
5. E. A. Allen, J. Del Gaudio and W. Wilkinson, *Thermochim. Acta* **11**, 197 (1975).
6. C. Sato and S. Hatakeyama, *Bull Chem. Soc. Jpn* **45**, 646 (1972).
7. A. Tatehata, T. Kumamaru and Y. Yamamoto, *J. Inorg. Nucl. Chem.* **33**, 3427 (1971).
8. J. M. Amigo, J. Garcia–Gonzales and C. Miravitlles, *J. Therm. Anal.* **3**, 169 (1971).
9. Z. Halmos and W. W. Wendlandt, *Thermochim. Acta* **4**, 25 (1972).
10. A. H. Verdonk, *Thermochim. Acta* **4**, 25 (1972).
11. V. V. Savant and C. C. Patel, *J. Less Common Metals* **24**, 459 (1971).
12. H. F. Steger, *J. Inorg. Nucl. Chem.* **34**, 175 (1972).

CHAPTER 22

Application to Space Studies

In considering the applications of DTA to Space Studies it is necessary to discuss two quite distinct areas of investigation:

(a) the use of DTA as an analytical tool in the investigation of properties of polymeric materials which may be used in the construction of a space vehicle.
(b) the design of DTA equipment which can be carried inside a space vehicle and will be used, at some time or another, to investigate materials obtained during the mission.

Since these applications are completely different they will be discussed independently.

INVESTIGATION OF THE PROPERTIES OF POLYMERIC MATERIALS FOR SPACE USE

We have already described, in Chapter 14, many of the ways in which DTA has been used to study polymeric materials and they will therefore not be covered here. However, before a material can be considered for use in the construction of any component in a space vehicle, it must be shown to fulfil a number of requirements. In normal satellite applications these can be summarized as:

(a) they should be stable under conditions of high vacuum ($< 10^{-6}$ Torr).
(b) they should be stable under conditions of extremes of temperature (< -150 to $+150$ °C) under high vacuum.
(c) they should be stable under conditions of exposure to radiation (i.r., u.v. particle).
(d) they should retain these properties in space for the duration of the mission (up to 7 years in some cases).

As can be imagined, this poses considerable problems for the materials engineer and not least among his worries is how to accelerate test methods to demonstrate that a material fulfills requirement (d). Differential thermal analysis can obviously be applied to temperature-stability measurements and is particularly useful in the determination of glass transition points, as has already been described.

Unfortunately, if these problems are not enough to worry about, the manned space programme led to even further difficulties. Two additional major requirements were placed on candidate materials:

(a) they should be non-flammable under the space vehicle atmosphere.
(b) they should not give off toxic fumes at the materials operating temperature.

The extent of the problem can be imagined when it is considered that in the Apollo and Skylab missions, the space vehicle atmosphere was 100% oxygen (at pressures ranging from 6.2 to 15.0 p.s.i. depending on the situation). The fact that polymeric materials were found and successfully used, which met these requirements, is a tribute to the NASA materials people. Fortunately the new generation of manned space vehicles, such as the Space Shuttle and Spacelab are planned to operate in a normal air atmosphere. However, the use of non-flammable materials still poses many problems.

The realization that the use of polymers incorporates many fire hazards has prompted much research into the stability of these materials under increased temperatures, and it is here that DTA can be an extremely important analytical tool.

If the burning process is considered on a micro scale, it is possible to make a classification into five quite distinct stages as follows:[1]

1. *Initial heating*, where the temperature of the polymer is gradually raised without any adverse effect being noticed on the polymer structure.

2. *Transition point*, where within a narrow temperature range the polymer ceases to be a relatively hard material and takes on a more rubbery condition. This transition point is the glass transition temperature.

3. *Initial degradation*; the temperature at which this occurs reflects the ease with which the least thermally stable bonds can be broken. This is frequently visible as a discolouration of the polymeric material. At this point, two properties of the material are particularly important. These are the decomposition temperature and heat of decomposition of the least stable bonds.

4. *Decomposition*, which may be defined as the failure of the polymer structure. This may lead to many different products being formed, ranging on one hand from complete disintegration (as in the case of polymethyl methacrylate) to the other extreme of rearrangement with little loss in weight. An example of this latter effect is in the rearrangement of polyacrylonitrile to give 'Black Orlon'.

When the temperature of failure of the least stable bonds is significantly lower than the temperature of breakdown of the general structure, then the degradation and decomposition stages are well separated. Unfortunately, it is frequently found that these two stages overlap.

Decomposition is greatly affected by the decomposition temperature and the heat of decomposition of the various bonds. If the process is endothermic heat must be supplied for the reaction to be sustained, whereas with an exothermic process enough heat is generated for the process to maintain itself. The mechanism of decomposition (that is the nature of the decomposition products) is also an important contributor to the overall process.

This stage of the reaction will produce two types of material—the polymer chain residue (carbonaceous) and polymer fragments, the latter being highly vulnerable to oxidation. Glowing of the carbonaceous residue can occur in the presence of oxygen, but the most probable source of ignition is in the gas phase in close proximity to the polymer residue.

5. *Oxidation* of the polymer fragments may produce heat and flame in the gas phase as well as glowing of the solid residue.

It can be shown that flame propagation rate (i.e. the speed with which a flame travels from the source of ignition to another point), heat of combustion and composition of polymeric materials are closely related. Those materials with high heats of combustion generally have a high flame propagation rate and vice versa. It is usually considered advisable to limit the use of plastics for space applications to those materials with a heat of combustion lower than c. 32.5 kJ g^{-1}.

In general terms, polymers may be made less flammable by the following methods.

1. Modification of the polymer so that decomposition yields non-combustible products, or that the gaseous products interfere with the exchange of combustion gases and air.
2. Modification to lead to a lower heat of combustion.
3. Modification to increase the amount of solid residue. This is very important in that structural integrity is retained and access of oxygen and heat is impeded. The use of these intumescent compounds is becoming important in space applications.[2]
4. Modification to increase decomposition temperature.

Unfortunately, the use of flame retardants often results in a product that produces toxic gases on smouldering and it is often necessary to have a balance between non-flammable and non-toxic materials.

It should now be apparent that DTA is extremely important in studies of materials from a flammability viewpoint. The heat of combustion can readily be determined from the area under the decomposition exotherm and the effects of flame retardants easily seen. Not only is the heat of combustion an important factor but the heat of decomposition of the polymer itself plays a major part in the overall process. It is preferable to use materials where the decomposition of the polymer, in an inert atmosphere, is endothermic, since this will at least be one step towards reducing the flammability of the material.

An example of the results obtained[3] on studies of polymers containing flame retardants is given in Fig. 22.1 which shows the effects of increasing concentration of tris–(2,3 dibromopropyl) phosphate on the decomposition of polystyrene. This particular compound appears to be an effective flame retardant according to mechanisms (2) and (3) above. When non-treated polystyrene is heated in nitrogen, a small endotherm appears between 176 and 256°C and a major one at

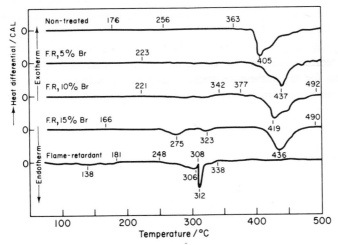

Fig. 22.1. (From Di Pietro and Stepniczka, Ref. 3.)

405°C. (Thermogravimetry shows that the first endotherm is due to the sublimation of unreacted monomer and low molecular weight polymers and that the second endotherm is due to decomposition.) Decomposition of the flame retardant form of polystyrene occurs at a somewhat lower temperature but the final product shows an increased char yield over the non-retardant system.

An example of the effect that a change in atmosphere may have on polymer decomposition is shown in Fig. 22.2 which shows the results obtained[4] on the decomposition of nylon 6.6 in nitrogen and in air. In air, an exothermic reaction occurs at about 185 °C followed by the endothermic melting peak at 250 °C. On further heating oxidation occurs leading to the exothermic peaks. By comparison, the trace obtained under nitrogen shows two endothermic peaks for melting at 265 °C and for decomposition over the range 350–450 °C.

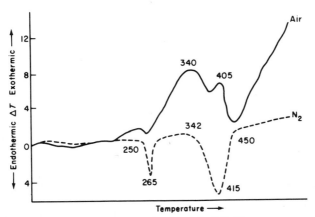

Fig. 22.2. (From Grzebieniak, Ref. 4.)

As can be seen, oxidation reactions can be identified simply by comparing the traces obtained in air and in nitrogen. The exotherm on a DTA trace has been used for studying the oxidation kinetics of materials.[5,6] A sample is held at constant temperature under an inert atmosphere; oxygen is then admitted and the time taken for the exotherm to appear is taken as the induction period. Experiments can be performed at a series of temperatures enabling the activation energy of the process to be determined.

It was mentioned earlier that the addition of flame retardants may well decrease flammability but at the expense of an increase in the production of toxic gases. Therefore many studies[7-9] have been carried out using DTA (and/or TG) coupled directly to either a gas chromatograph or mass spectrometer. This enables a direct estimation to be made of amounts of toxic products evolved from a degradation process. It is certainly true that this information is essential before full evaluation of the effects of flame retardants on polymer systems can be made.

DTA EQUIPMENT SUITABLE FOR USE DURING A MISSION

The design of a differential thermal analyser involves consideration of many experimental variables and these have already been described. However, for use in space, or as part of the payload of a space vehicle, many more restraints have to be imposed on the design. Among the more important of these are the following:

1. Robust construction; able to withstand the vibration and acceleration forces involved on lift off.
2. Small size; necessary because of the very high cost of every 1 lb of payload.
3. High speed; in other words it should be able to operate at high sampling speeds, heating rates etc. It should also be either fully automatic even to the extent of self-sampling, or, at least in a manned mission, it should require minimum attention.
4. It should be capable of providing an output which can either be stored on inboard computers, or be transmitted to earth for processing.
5. It should have low power consumption.
6. It should be constructed of materials which meet all the requirements given at the beginning of this section.

One immediate conclusion that can be drawn from these restraints is that no 'off the shelf' equipment can be considered for such applications. It would also be true to say that it is doubtful if DTA will ever seriously be considered as an essential part of a spacecraft payload. This is not to say that DTA is not suited to the analysis of, say for example, moon rocks, but as should be clear by now, DTA is not a specific technique. Differential thermal analysis is at its best when used in conjunction with other techniques and only in few cases is DTA used on its own as an analysis tool. Because of the payload costs on space projects it is by far preferable to include specific equipment such as the combined gas chromato-graph/mass spectrometer which was used on the Viking probe to Mars in 1975.

Design studies have been made[10] on suitable DTA equipment which could possibly meet the requirements given above. This design was proposed for use on the Mariner missions to Mars, but as far as is known the equipment was never flown. However, the following description of the equipment will serve to show just how far it is possible to adapt DTA to meet the specific needs of the space industry.

Figure 22.3 shows the arrangement of the equipment together with an effluent gas lead out. The first thing that is apparent is the very small size of the system,

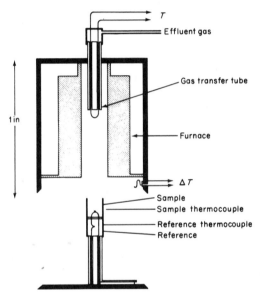

Fig. 22.3. DTA equipment with an effluent gas collection enclosure. (From Bollin, Ref. 10.)

fitting into a cube of dimensions of 1 in on all sides. A major modification to the design of the head assembly is that the normal bilaterial symmetry has been replaced by linear–axial symmetry, thus enabling a much smaller overall unit to be produced. A further reason for such a system is that automatic sampling is made much easier; the reference material is prepacked and therefore remains unchanged throughout the life of the equipment. The fact that such a system is capable of operating at high heating rates is demonstrated in Fig. 22.4, where the DTA traces for a series of desert soils were obtained at a heating rate of 80 °C min⁻¹. It is stated that heating rates of up to 2400 °C min⁻¹ are possible and that base-line stability is reasonable. However, owing to peak overlap this is not a practical proposition and heating rates in the range 50–100 °C are more reason-able.

Curve	mg	Soil No	Location	Total organic matter %
A	20	I-2	Thermal, Calif.	0.4
B	18.5	297	Egypt, Eastern Sahara	0.2
C	17	291	Chili, Atacama	0.06
D	20	76-1	Mojave Desert	0.9

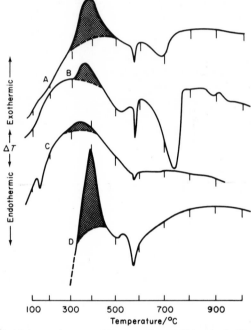

Fig. 22.4. Moderately high-speed DTA of desert soils at 80°C min⁻¹. (From Bollin, Ref. 10.)

The applications of such a DTA system to the analysis of mineralogical samples with a view to the determination of planetary environmental parameters has also been described.[11]

REFERENCES

1. C. J. Hilado, *Flammability Handbook for Plastics*, Technomic Publishing Co. Inc., Stamford, Conn., U.S.A., 1969.
2. G. M. Fohlen, J. A. Parker, S. R. Riccitiello and P. M. Sawke, *Conference on Improved Fire Safety*, Houston, Texas 1970, NASA SP-5096, 1971, p. 111.
3. J. Di Pietro and H. Stepniczka, *J. Fire Flamm.* **2**, 36 (1971).
4. K. Grzebieniak, *Polbm. Tworz. Wlelk* 336 (1973).
5. B. Ke, *Newer Methods of Polymer Characterization*, Interscience Publishers, Chichester, Sussex, U.K. 1964.

6. A. Rudin, *Ind. Eng. Chem.* **53**, 137 (1961).
7. E. A. Boetner, *J. Appl. Polym. Sci.* **13**, 377 (1969).
8. G. P. Shulman and H. W. Lochte, *J. Appl. Polym. Sci.* **10**, 619 (1966).
9. W. D. Wooley, *Plast. Polym.* 280 (1973).
10. E. M. Bollin, *Thermal Analysis, Proceedings of the Second International Conference for Thermal Analysis*, Worcester, U.S.A. 1968, Academic Press, New York, 1969, p. 255.
11. E. M. Bollin, *Thermal Analysis, Proceedings of the Second International Conference for Thermal Analysis*, Worcester, U.S.A. 1968, Academic Press, New York, 1969, p. 1387.

CHAPTER 23

Reaction Kinetics

INTRODUCTION

This chapter on kinetic data from DTA has been deliberately left to the end of this book for many reasons. By now the reader should be well aware of the many problems associated with the sensible interpretation of the results obtained from a DTA trace: the problems associated with the sample, namely effects of particle size, sample size (volume, depth and uniformity of packing); those associated with the sample environment, namely crucible shape, gaseous atmosphere and so on. There is no one area of study more affected by these variables than in the determination of kinetic parameters by DTA. Unfortunately this fact of life seems to have been glossed over by many workers and this is evidenced by the many scientific papers on the subject which contain results of little value and even less meaning.

Let us first consider some general aspects which may help in clarifying the above statements. The initial study of Borchardt and Daniels[1] was an elegant piece of theoretical and experimental work. However, as will be described later, although theoretically correct for solution kinetics, certain of the assumptions made render the treatment inapplicable to solid state kinetics. In fact, these incorrect assumptions are the basis of the failure of many reported investigations on solid reactions and it is therefore important that these be clarified further.

The first assumption is that the temperature of the sample is, at all times, uniform. This is obviously incorrect for solid state reactions. The major problem with any thermoanalytical technique involving a continually changing temperature is that a temperature gradient will always be present within the sample. This effect, will of course, be magnified as the heating rate is increased. Even small differences in temperature will lead to marked changes in the rate of reaction. (Even heat liberated or absorbed by the reaction itself will cause the temperature of the reaction interface to differ from that of its surroundings.) The sample at the centre of the crucible can only receive heat from the furnace through the sample close to the crucible walls. If the reaction is occurring at a moderate rate the sample near the crucible wall may well absorb more heat than it passes on. To reduce these effects, slow heating rates, or small sample weights must be used. Of these two alternatives, the former is not recommended since the small values of ΔT obtained will be too low to measure accurately. The use of small samples is more acceptable and will often yield more precise information. However, care must be exercised when using small samples as will be discussed later.

The second problem area is encountered in the assumption that the heat capacity of the system will remain constant throughout the course of the reaction. Furthermore, it is assumed that the heat capacities of the sample and reference material are the same. This is obviously not true since the sample will undergo many transformations during the course of reaction; transformations involving weight changes, phase transitions, sintering and so on. The changes in heat capacity of the sample during the course of a reaction can clearly be seen by changes in the slope of the base line on the DTA trace. Vold[2] has suggested that these effects may be minimized by ensuring that the heat capacity of the sample holder should be large. Unfortunately this approach, although reasonable on theoretical grounds, is untenable experimentally since the use of massive sample holders leads to a marked decrease in sensitivity.

These two problem areas therefore illustrate the main difficulties associated with the applications of DTA to kinetic calculations. The situation is made more complicated still by two basic misunderstandings on the part of many workers in this field. The following comments are not solely directed at the worker using DTA but are equally as applicable to the user of the sister technique, TG.

The first misunderstanding is that reactions obey equations. It is a valid argument that if you manipulate the experimental data sufficiently, or if you take an equation with enough terms in it, you can satisfactorily describe any chemical process. The problem encountered with the first approach is that there is a natural human tendency to try and make the data fit an equation which would seem to be the correct equation initially assumed by the worker concerned. In many cases, the experimenter already has a preconceived idea of how the reaction should behave and this will be reflected in that he will attempt to fit his data to his assumed process. Even if the data only fit the equation over 10% of the process range, then this is often considered sufficient, and the remaining 90% is ignored.

The second misunderstanding is that reactions will always obey a theory. This is obvious nonsense and if true would halt any further advances in science. If the experimental data, after a full investigation of the experimental variables, is shown to be reliable and reproducible and the present theory cannot be applied, then the theory must be modified. This, of course, is not to say that the worker should not attempt to fit his experimental data to the well established equations. But what he should not do is to ignore any deviations that show up in the fit between experimental results and theory, simply in order to report an acceptable result.

THEORETICAL APPROACH TO KINETIC DATA FROM DTA

This section will consider several equations, and their derivations, which have been proposed for the determination of kinetic data from DTA. Inherent in the treatment of these is an understanding and knowledge of the basic equations for solid state reactions. It is not within the scope of this section to expand on these equations and therefore the reader is advised to consult one or more of the available reviews on the subject.[3-7]

Method of Borchardt and Daniels[1]

This work was the first which attempted to put the determination of kinetic parameters from a DTA trace onto a sound theoretical footing. They studied the decomposition of a 0.4 M solution of benzenediazonium chloride in water and derived an expression which relates the rate constant to several parameters of a DTA peak. A detailed derivation of this expression will be given because a full understanding of the assumptions made is essential in order to realize the limitations of the theory. A schematic diagram of the apparatus used is shown in Fig. 23.1. It should be noted that the authors well appreciated the problems of

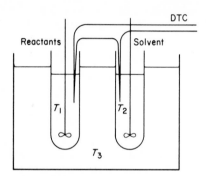

Fig. 23.1. Differential thermal analysis apparatus for obtaining kinetic data for reactions occurring in solution. (Borchardt and Daniels, Ref. 1.)

thermal gradients existing within the sample, since the sample and reference solutions and thermostat bath are stirred. It should also be recorded that strictly, their approach is applicable only to reactions in solution (although this does not appear to have registered with many workers). They propose that the following equations represent the heat balance between the sample and the reference material.

For the sample

$$Cp_S dT_S = dH + K_S(T_B - T_S)dt \tag{1}$$

and for the reference

$$Cp_R dT_R = K_R(T_B - T_R)dt \tag{2}$$

where Cp is the heat capacity of the sample (Cp_S) and reference (Cp_R), dT is the temperature change of the sample (dT_S) and reference (dT_R), K is the heat transfer coefficient of the sample (K_S) and reference K_R, and dH is the heat change in time dt, and T is the temperature of the sample (T_S), reference (T_R) and thermostat bath (T_B.)

Three assumptions are now made in order to simplify the mathematical treatment:

1. Since identical cells are used and are filled to the same level, $K_S = K_R$ and can be replaced by a constant K.
2. The temperature is assumed to be uniform throughout the sample and the reference material. This condition can be satisfied for stirred liquids but, as has already been mentioned, it cannot be satisfied for solids.
3. It is considered that a 0.4 M solution of benzenediazonium chloride in water has the same heat capacity as that of pure water, i.e.

$$Cp_R = Cp_S = Cp$$

Making these assumptions and taking Eqn (2) from Eqn (1) gives

$$Cp(dT_S - dT_R) = dH - K\Delta T dt \qquad (3)$$

where $\Delta T = T_S - T_R$.

Rearranging (3) yields

$$dH = Cp(dT_S - dT_R) + K\Delta T dt \qquad (4)$$

This gives the heat change over time, dt, and thus to determine the total heat change, Eqn (4) must be integrated over limits $t = 0$ to $t = \infty$, i.e.

$$\Delta H = Cp(\Delta T_\infty - \Delta T_0) + K\int_0^\infty \Delta T dt \qquad (5)$$

When $t = 0$, $\Delta T = 0$, and when $t \infty =$, $\Delta T = 0$, so that the integral $\int_0^\infty \Delta T dt$ is represented by the total area A under the curve given in Fig. 23.2, i.e.

$$\Delta H = KA \qquad (6)$$

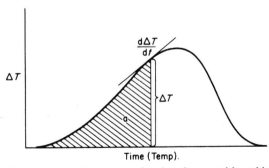

Fig. 23.2. Differential thermal analysis curve showing the quantities which are measured in order to evaluate the rate constants for the reaction giving rise to the curve. (Borchardt and Daniels, Ref. 1.)

The reader will have met this equation before in that it is used as a basis for many applications of quantitative DTA (see Chapter 4). If it is assumed that ΔH is the total heat evolved in the reaction of x_0 moles of substance, then, at any moment, dH is proportional to dx, the amount reacted, i.e.

$$-\frac{dH}{\Delta H} = \frac{dx}{x_0} \tag{7}$$

or

$$dH = -\frac{(KA)dx}{x_0} \tag{8}$$

Substitution of Eqn (8) into Eqn (3) and differentiating with respect to t gives

$$-\frac{dx}{dt} = \frac{x_0}{KA}\left[Cp\frac{d\Delta T}{dt} + K\Delta T\right] \tag{9}$$

Eqn (9) describes the rate of reaction at any temperature in terms of the slope $d\Delta t/dt$ and height ΔT of the DTA curve at that temperature.

The number of moles present at time t is equal to the number of moles originally present less the numbers of moles reacted, i.e.

$$x = x_0 - \int_0^t -\left(\frac{dx}{dt}\right)dt \tag{10}$$

Substitution of this into Eqn (9) and integrating gives

$$x = x_0 - \frac{x_0}{KA}\left(Cp\Delta T + Ka\right) \tag{11}$$

where a is the area enclosed by the DTA peak at time t. A particular model for the kinetic process is now assumed. For a reaction in solution it is not unreasonable to suppose that the order of reaction equation will be followed, i.e.

$$-\frac{dc}{dt} = kc^n \tag{12}$$

where c is the concentration of reactant, k is the rate constant, and n is the order of reaction.

If Eqn (12) is rewritten in terms of the number of moles of reactant in volume V, namely

$$-\frac{dx}{Vdt} = k\left(\frac{x}{v}\right)^n \tag{13}$$

and this is substituted into Eqn (9) and (11) we finally arrive at the general expression:

$$k = \left(\frac{KAV}{x_0}\right)^{(n-1)}\left[\frac{\frac{Cpd\Delta T + K\Delta T}{dt}}{(K(A-a) - Cp\Delta T)^n}\right] \tag{14}$$

where V is the volume of the solution containing the reactants. For the special case of first-order kinetics where $n = 1$, this simplifies to:

$$k = \frac{\overline{Cpd\Delta T + K\Delta T}}{K(A-a) - Cp\Delta T} \tag{15}$$

It is possible to simplify the equation still further for first-order kinetics, if it is assumed that the term $Cpd\Delta T/dt$ is small compared to $K\Delta T$ and that $Cp\Delta T$ is small compared to $K(A-a)$. If these assumptions are made then Eqn (15) simplifies to

$$k = \frac{\Delta T}{(A-a)} \tag{16}$$

The use of this equation, however, should be avoided without a full consideration as to whether or not the above assumptions are valid.

The general equation of Borchardt and Daniels has been extended by Sharp[8] to include diffusion-controlled reactions. For the detailed theory given above to be applicable to solid state reactions, it is implicit that all the assumptions made must be valid. Unfortunately this is not so and Borchardt and Daniels themselves conclude that, although when applied to solution kinetics the treatment is valid, it is not so for solid state reactions. For a detailed critique of this method the reader is recommended to read the excellent review by Garn.[9]

Method of Peron and Mathieu[10]

This is an extension of the expression derived by Freeman and Carroll[11] for the determination of kinetic parameters from TG studies. The generalized equation for the dependence of the rate constant on temperature is:

$$k = A_F \exp\left(\frac{-E}{RT}\right) \tag{17}$$

where A_F is the pre-exponential factor (given here a subscript F to avoid confusion with the area term A), E is the activation energy, and R is the gas constant.

If this is combined with Eqn (14) we have

$$\ln A_F - \frac{E}{RT} = \frac{(n-1)\ln(KAV/x_0) + \ln(Cpd\Delta T/dt + K\Delta T)}{-n\ln(K(A-a) - Cp\Delta T)} \tag{18}$$

Differentiation of this equation with respect to T gives

$$\frac{E\,dT}{RT^2} = n\,d\ln\left[K(A-a) - Cp\Delta T\right] - d\ln(Cpd\Delta T/dt + K\Delta T) \tag{19}$$

Rearrangement and integration of (19) gives

$$-\frac{(E/R)\Delta(1/T)}{\Delta\ln[K(A-a) - Cp\Delta T]} = -n + \frac{\Delta\ln(Cpd\Delta T/dt + K\Delta T)}{\Delta\ln[K(A-a) - Cp\Delta T]} \tag{20}$$

This equation is therefore in such a form that a linear plot should be obtained, n being given by the intercept and E by the slope. Unfortunately this approach suffers from the same failing as does the Freemann Carroll expression in that it is a difference–difference technique. That is to say that it is necessary to determine the difference between successive differences in values of $\Delta T(d\Delta T)$. Thus an error in one value of ΔT will affect not only one value of the Perron Mathieu Plot, but will affect values either side of it. It is therefore to be expected that the points will show a considerable scatter and the value of the order of reaction so determined will be subject to appreciable error.

Method of Kissinger[12, 13]

Kissinger[12] has proposed that the variation of the peak temperature of a DTA peak with heating rate may be used to determine the activation energy of first-order processes. Later work[13] extended this approach to cover reactions of any order. Kissinger also proposed [13] that information on the kinetics of a process, with particular reference to determination of n, could be obtained from the shape of the DTA trace. This was termed the 'shape index'. The whole basis of Kissinger's study is that the temperature of maximum deflection in a DTA peak is also the temperature at which the reaction rate is at a maximum. This assumption has been shown to be incorrect;[14] the maximum rate of reaction occurs somewhere before this point. Because of this fact, it is doubtful whether this method can be applied with any meaning to solid state reactions. Garn[15] has published a detailed analysis of this approach. Several modifications to Kissinger's method have been proposed with particular reference to DSC. Of special relevance is the work of Roger and Smith[16] and Ozawa.[17, 18]

EXPERIMENTAL CONSIDERATIONS

The factors listed below are those to be considered in addition to the experimental variables discussed in Chapter 3. It cannot be emphasised enough that before any experimental data is subjected to kinetic analysis its validity must first be evaluated.

Apparatus effects

A thermocouple only measures its own temperature. This at first sight may seem to be a statement of the obvious but it does have a definite impact as far as kinetics is concerned. In conventional DTA equipment a thermocouple is placed in contact with the sample crucible and not with the sample itself. In other words, the temperature that the thermocouple registers will depend on the amount of heat transmitted from the sample through the crucible. This leads to the correct conclusion that the shape of the sample assembly, with respect to sample block and crucible, can affect the data obtained. For instance, if a crucible is used which has low sides in relation to the base area and is sitting in a ceramic block with conventional thermocouple placement, the rate of heating of the sample will not be uniform. The sides of the crucible will be heated up more quickly than the base and hence the reaction will occur more quickly at the sample in contact with the crucible wall. If a fast heating rate is used then the situation will be even worse. In consideration of this fact it must be remembered that it is normal practice to use conventional DTA crucibles for kinetic work, where sample blocks are necessary. More recent designs have eliminated the sample block, and flat pan crucibles resting on disc thermocouples are used. This situation is better, in that small samples can be used, but has the disadvantage that accurate and reproducible positioning of the crucible onto the thermocouple cannot be guaranteed.

Sample effects

Wiedemann[19] has stated that uniform packing density and thickness are essential to obtain reproducible results when determining activation energies. In the case where one product of a reaction is a gas, this gas must pass through a bed of reacted material, so if this is not an instantaneous process, some pressure of gas will exist at the reaction interface. This will tend to slow down the reaction and thus will act in opposition to the increase in rate brought about by any increase in temperature. In fact the variables described earlier concerning DTA samples should be considered. The reader would be well advised to control these parameters as far as possible, in order to fix some of these variables. For instance, the use of a constant sample weight with a constant particle size and packed into the crucible under controlled conditions would be a useful starting point.

EFFECTS OF KINETIC PARAMETERS ON THE SHAPE OF THE DTA PEAK

Kissinger[12] was one of the first to point out the marked dependence of the shape of a DTA peak on the reaction order. As has been mentioned earlier, he attempted to use his 'shape index' as a first approximation in the determination of n. In general an increase in activation energy will decrease the size of a DTA peak whereas an increase in the pre-exponential factor will have the converse effect. Each of these produces a negligible change in the shape of the peak. The effect of an increase in n is to decrease somewhat the size of the peak, but with a correspondingly drastic change in shape. These considerations are well demonstrated by the work of Reed.[14]

EXAMPLES OF SOME KINETIC INVESTIGATIONS

It would be impracticable to detail large numbers of reactions which have been studied by DTA. Some examples are given, which are considered by the authors to be of interest. Further examples may be found by consultation of *Thermal Analysis Abstracts*[20] or similar abstract journals.

Perhaps the most widely reported study of kinetics by DTA is that of the thermal decomposition of benzene diazonium chloride. Borchardt and Daniels[1] studied the exothermic peak on the DTA trace obtained by heating 35 ml of a 0.4 M solution from 2 °C to 70 °C at a rate of rise of temperature of 1 °C min^{-1}. The rate of reaction was calculated assuming the order of reaction (n) had values of 0, 1, 2 and 3. In the case where $n = 1$, a linear activation energy plot was obtained (Fig. 23.3) giving a value of the activation of energy of 28.6 kcal mol^{-1}. The results obtained agreed well with isothermal studies, where first-order kinetics was found with an activation energy of 27.2 kcal mol^{-1}. Subsequent experiments by Reed[14] confirmed the validity of Borchardt and Daniels's earlier work. It is somewhat surprising to find that there is very little subsequent work published in the literature on the uses of DTA in the study of solution

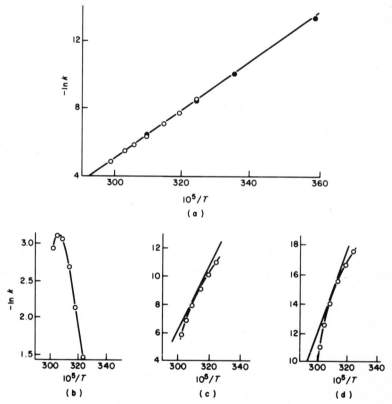

Fig. 23.3 (a) Arrhenius plot for the decomposition of benzenediazonium chloride assuming $n = 1$. (b) (c) and (d) Arrhenius plots for the decomposition of benzenediazonium chloride assuming $n = 0$, $n = 2$ and $n = 3$, respectively. Temperature in K; (Borchardt and Daniels, Ref. 1.)

kinetics. The kinetic treatment of Borchardt and Daniels is valid theoretically for this type of investigation and would therefore appear to give kinetic parameters rapidly, that are sufficiently accurate for most workers.

Thomas and Clarke[21] have attempted to use a modified version of the Borchardt and Daniels equations for studies of the dehydration of manganous formate dihydrate. From considerations of the heat effects involved in a DSC experiment they derive the general expression

$$\frac{d\alpha}{dt} = \frac{1}{A}\frac{dH}{dt} \tag{21}$$

Their principle assumption is that the enthalpy change in a small interval of time dt is directly proportional to the number of moles that react in that time interval.

If the linear region of the kinetic process is considered the relationship between the rate constant and the fraction of sample decomposed is given by

$$\frac{d\alpha}{dt} = k \text{ (from } \alpha = kt) \tag{22}$$

to give

$$k = \frac{1}{A}\frac{dH}{dt} \tag{23}$$

This expression was tested on the dehydration of manganous formate dihydrate, since it has been demonstrated in earlier work[22] that the process proceeds via interfacial growth of the dehydration product. This was reflected in the isothermal TG plots which showed distinct regions where the weight loss was rectilinear with time, over the α range 0.2–0.5. Therefore it would be expected that Eqn (22) would be valid for the DSC trace over the same range of α. This was in fact shown to be the case, as is illustrated in Fig. 23.4 which is a plot of log dH/dt v. $1/T$ over the α range 0.19–0.46.

The circular points refer to the DSC data whereas the crossed points were obtained by isothermal TG. It is also interesting to note that the value of the activation energy obtained, 16.5 ± 0.5 kcal mol^{-1}, is unaffected if the heating rate is changed from 8 °C min^{-1} to 16 °C min^{-1}.

An interesting application[23] of the methods of Kissinger[12] and Ozawa[18] is illustrated in studies of the lattice reorganization of copper. A pronounced peak symmetry is displayed on the DSC trace under different heating rates and it is shown that the temperature of half recrystallization ($\alpha = 0.5$) coincides with the peak temperature T_M. Figure 23.5 illustrates the plots obtained using the methods of Ozawa and Kissinger, showing the dependence of peak temperature T_M on heating rate ϕ. The solid circles represent corrected values of the temperature as determined by separate melting point calibration points. This corrected line yields an activation energy some 25% higher than the uncorrected value and more closely agrees with the value determined by isothermal measurements.

It was earlier pointed out that the method of Kissinger is not valid for DTA, since it is based on the incorrect assumption that the peak temperature of a DTA

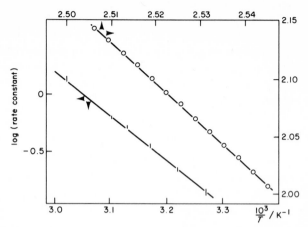

Fig. 23.4. Arrhenius plots constructed from DTA data. (From Thomas and Clarke, Ref. 21.)

peak corresponds to the maximum rate of reaction. Ozawa,[18] concludes that the assumption is valid for DSC curves because of their derivative nature. This, however, is the subject of some controversy, which lies outside the scope of this book. Nevertheless, even if this assumption is valid, the position of the thermoelectric sensors outside the sample and their consequent thermal lag, which increases with heating rate, is an intrinsic characteristic of DSC equipment. Therefore Kissinger's and Ozawa's plots tend to underestimate the value of the activation energy, if a calibration of sample temperature is neglected. This error may be as much as 30% in cases of materials with low thermal conductivity.

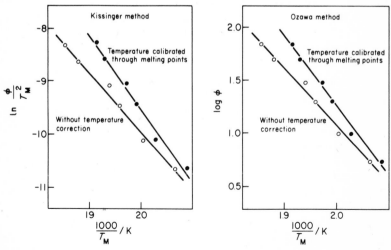

Fig. 23.5. Kissinger and Ozawa. Plots for the lattice reorganisation of copper. (From Lucci and Tamanini, Ref. 23.)

The decomposition of sodium hydrogen carbonate is a good illustration of the dependence of the kinetic parameters on sample mass. The results obtained by Guarini et al.[24] are illustrated in Fig. 23.6 where the logarithm of the $n = 1$ rate constant of the standard equation $d\alpha/dt = k(1 - \alpha)^n$, deduced at constant heating rate, is plotted against $1/T$. The plots become progressively more curved as the sample mass increases. This yields activation energies which are different for different ranges of α and agree with the literature value only in a very restricted part of the curve. These authors have used an extrapolation method based on

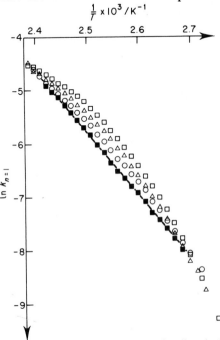

Fig. 23.6. Activation energy plots for the decomposition of sodium hydrogen carbonate, using different masses. (From Guarini et al., Ref. 24.)

plotting the fractional decomposition at a given temperature, as a function of mass, using samples of different mass.

Within the limits of error these plots were linear and enabled a value of the limiting fractional decomposition at 'zero mass' to be calculated. A similar procedure allowed the height of the peak (proportional to $d\alpha/dt$) at $m = 0$ to be computed. It was therefore possible to obtain a value of $k_{m = 0}$ from the equation

$$k_{m = 0} = \frac{\lim\limits_{m \to 0} \dfrac{d\alpha}{dt}}{\lim\limits_{m \to 0} (1 - \alpha)}$$

with $n = 1$.

The Arrhenius plot for log $k_{m=0}$ versus T^{-1} is also shown in Fig. 23.6 and, as is apparent, a good straight line fit is obtained.

The valuation of the activation energy obtained agrees well with the literature value and is valid over a large range of a. It would therefore appear that, certainly from a mathematical viewpoint, the method is valid. However, what physical significance can be placed on $a_{m=0}$ and $k_{m=0}$ is a matter of conjecture.

A process that has been widely studied is the dehydroxylation of Kaolinite. The reaction has been variously described as following the order of reaction mechanism, with n having a value of 1 [12, 25, 27] or 3 [27], or being diffusion controlled [28-30]. Most of the earlier studies [12, 25-27] have taken place using air as the atmosphere surrounding the sample. Problems involved in the effects of the presence of water vapour around the sample have not been considered and therefore it is doubtful whether these results are valid. It is, in fact, interesting to note that in the examples where the atmosphere around the sample is water vapour at reduced pressure, the reaction is found to be diffusion controlled.[28-30] Similar variations in the value of the activation energy are found even where the same theoretical model is used. This variation too can, to a large extent, be explained in terms of the effects due to the sample atmosphere. However, it is probable that these differences result from a lack of control of the sample parameters, namely sample pretreatment, differing sample localities and so on.

In conclusion, let us consider what are the main reasons given by workers in kinetics, for using dynamic thermal methods.
Two reasons are normally quoted:

1. Dynamic methods follow the kinetics over the whole temperature range.
2. In isothermal methods some reaction occurs before the temperature of interest is reached.

Sharp,[8] in his review on kinetics from DTA, very succinctly puts these views into perspective by stating that, rather than justifying dynamic methods, these are simply two failings of isothermal methods. It is hoped that the reader is by now convinced that the limitations of dynamic methods, when applied to solids, are far more serious.

REFERENCES

1.   H. J. Borchardt and F. Daniels, *J. Am. Chem. Soc.* **79**, 41 (1957).
2.   M. J. Vold, *Anal. Chem.* **21**, 683 (1949).
3.   D. Dollimore and C. J. Keattch, *Introduction to Thermogravimetry*, 2nd Edn, Heyden and Son Ltd., London, 1975, p. 57.
4.   D. A. Young, *Decomposition of Solids*, Pergamon Press, Oxford, 1966.
5.   W. E. Garner, *Chemistry of the Solid State*, Butterworth, London, 1955.
6.   B. Delmon, *Introduction a la Cinétique Hétérogène*, Editions Technip, Paris, 1969.
7.   H. Schmalzried, *Solid State Reactions*, Verlag Chemie, Weinhein, 1974.

8. J. H. Sharp, *Differential Thermal Analysis*, (R. C. MacKenzie, Ed.), Academic Press, New York, 1973, p. 47.
9. P. D. Garn, CRC Critical Reviews in Analytical Chemistry 65 (1972).
10. R. Perron and A. Mathieu, *Chem. Anal.* **46**, 293 (1964).
11. E. S. Freeman and B. Carroll, *J. Phys. Chem.* **62**, 394 (1958).
12. H. E. Kissinger, *J. Res. Nat. Bur. Stand.* **57**, 217 (1956).
13. H. E. Kissinger, *Anal. Chem.* **29**, 1702 (1957).
14. R. L. Reed, B. S. Gottfried and L. Weber, *Ind. Eng. Chem. Fundls.* **4**, 38 (1965).
15. P. D. Garn, *Thermoanalytical Methods of Investigation*, Academic Press, New York, 1965.
16. R. N. Rogers and L. C. Smith, *Thermochim. Acta* **1**, 1 (1970).
17. T. Ozawa, *Bull. Chem. Soc. Jpn* **38**, 1881 (1965).
18. T. Ozawa, *J. Thermal Anal.* **2**, 301 (1970).
19. H. G. Wiedemann, A. van Tets and P. H. Vaughan, *Pittsburgh Conference on Analytical Chemistry and Applied Spectroscopy*, 1966.
20. *Thermal Analysis Abstracts*, published quarterly by Heyden and Son Ltd., London.
21. J. M. Thomas and T. A. Clark, *J. Chem. Soc.* (A), 457 (1968).
22. R. C. Eckhardt and T. Flanagan, *Trans. Faraday Soc.* **60**, 1289 (1964).
23. A. Lucci and M. Tamanini, *Thermochim. Acta* **13**, 147 (1975).
24. G. G. T. Guarini, R. Spinicci, F. M. Carlini and D. Donati, *J. Thermal Anal.* **5**, 307 (1973).
25. E. B. Allison, *Clay Miner. Bull.* **2**, 242 (1955).
26. T. Jacobs, *Nature (London)* **182**, 1086 (1958).
27. J. N. Weber and R. Roy, *Am. Mineral.* **50**, 1038 (1956).
28. J. B. Holt, I. B. Cutler and M. E. Wadsworth, *J. Am. Ceram. Soc.* **45**, 133 (1962).
29. B. N. N. Achar, G. W. Brindley and J. H. Sharp, *Proceedings of the International Clay Conference*, Vol. 1, (L. Heller and A. Weiss, Eds.), 1966, p. 67.
30. G. W. Brindley, B. N. N. Achar and J. H. Sharp, *Am. Mineral.* **52**, 1697 (1967).

Index

DATE DUE

GAYLORD			PRINTED IN U.S.A.